Nectar and Pollen Plants of Oregon and the Pacific Northwest

An illustrated dictionary of plants used by honey bees

by

D.M. Burgett, B.A. Stringer and L.D. Johnston

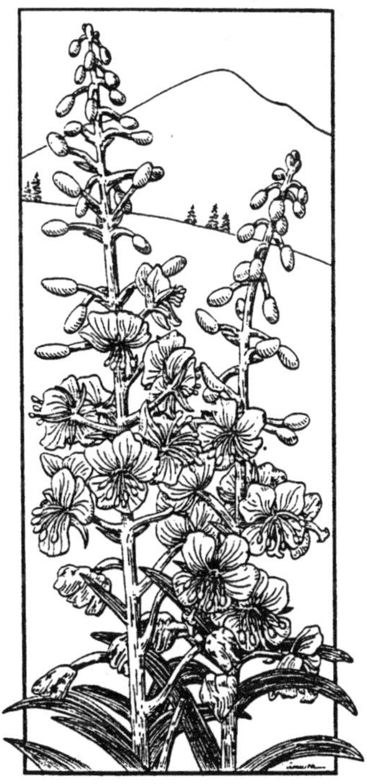

Honeystone Press
Blodgett, Oregon
1989

Cover; Fireweed, *Epilobium angustifolium*, a premier honey plant of the Pacific Northwest. Art and design by Anna Stone Asquith.

Published by Honeystone Press, PO Box 511, Blodgett, Oregon 97326, USA.
(1989)

CONTENTS

ACKNOWLEDGEMENTS

Grateful acknowledgement is made to the following for permission to reproduce illustrations: Special thanks to La Rea Johnston for use of illustrations from the *Handbook of Northwestern Plants* and *Gilkey's Weeds of the Pacific Northwest*; University of California Press for use of illustrations from *A Manual of Flowering Plants of California*; Stanford University Press for use of illustrations from the *Illustrated Flora of the Pacific States*; United States Department of Agriculture Forest Service for use of illustrations from the *Key to Important Woody Plants of Eastern Oregon and Washington* and *Forest Trees of the Pacific Slope*; University of Washington Press for use of illustrations from *Vascular Plants of the Pacific Northwest*; Dover Publications for use of illustrations from *Manual of the Trees of North America*; Timber Press, Inc. for use of illustrations from the *Manual of Cultivated Broad-leaved Trees and Shrubs*; Oregon State University Extension Service for use of illustrations from various of their publications; Emile T. Labadie for use of illustrations from *Ornamental Shrubs for use in the Western Landscape*; A.I. Root Co. for use of illustrations from 1901 *ABC and XYZ of Bee Culture*; the Royal Botanic Gardens at Kew for information regarding *The Illustrated Dictionary of Gardening*; and the Macmillan Publishing Company for information regarding illustrations in the *Standard Cyclopedia of Horticulture* and the *Cyclopedia of American Agriculture*.

We wish to acknowledge the assistance of Ray Drapek, Entomology Department, Oregon State University, for computer generation of Map 2.

The authors also wish to thank those beekeepers from the major beekeeping regions of Oregon for early review of the manuscript.

Finally sincere thanks to Marshall Dunham of Honeystone Press, without whose patience and support this book could not have been written.

ILLUSTRATION CREDITS

Reprinted with permission of La Rea D. Johnston from **Handbook of Northwestern Plants** by Helen M. Gilkey and La Rea J. Dennis. Copyright 1980 by La Rea J. Dennis: Figures 10, 11, 12, 17, 23, 30, 33, 34, 37b&c, 41, 44, 46, 53, 57, 65, 68, 73, 75, 77, 78, 84, 85, 86, 87, 89, 91, 106, 109, 112, 114, 117, 118, 123, 130, 133, 134, 136, 137, 142, 143, 155, 157, 158, 160, 168, 169, 174, 177, 178, 179, 180, 188, 191, 192, 194, 199, 203, 204, 206, 208, 210, 211, 212, 215, 216, 217, 219, 226, 229, 231, 234, 235, 239, 241, 249.

Reprinted with permission of La Rea D. Johnston from **Gilkey's Weeds of the Pacific Northwest** by La Rea J. Dennis. Copyright 1980 by La Rea J. Dennis: Figures 16, 19, 20, 29, 47, 55, 59, 70, 71, 72, 74, 76, 83, 95, 99, 104, 110, 111,

NECTAR AND POLLEN PLANTS OF OREGON
and the Pacific Northwest
An illustrated guide to plants used by honey bees

Preface

In the compilation of this book, the 1942 Oregon State College bulletin by Scullen and Vansell has been updated and supplemented by comments from beekeepers representing the primary beekeeping areas of Oregon. Because state lines do not dictate botanical boundaries, the influence of this work extends beyond the state of Oregon into other regions of the Pacific Northwest. Oregon's geographic diversity supports plants representing each of the neighboring states. Every plant entry in this guide is illustrated, and flower color noted, to simplify identification.

Introduction

Honey bees obtain almost all of their nutritional requirements from nectar and pollen produced by flowers. It therefore benefits the beekeeper to recognize those plants which provide forage for bees. The availability of nectar and pollen producing flora limits the number and location of honey bee colonies in any area, and the quantity and quality of the honey crop depends largely upon the pasturage within several miles of an apiary location. Ideal sites provide a succession of blooming plants throughout the active foraging season, but these areas have become fewer due to changing agricultural plantings and farming methods, urbanization and pesticidal practices. Nearly all areas of Oregon can sustain the few colonies of a hobbyist through the active foraging season, but commercial honey producers must move their colonies to take advantage of the most productive sites.

A colony of honey bees is not a static organism. It increases and decreases in size in a predictable manner which is closely allied with the seasons. During the active brood rearing season a colony requires a daily intake of nectar and pollen to meet the nutritional needs of the brood and adult bees. If insufficient forage is available from the botanical environment, a colony must utilize stored honey and pollen. On the other hand, nectar and pollen collected in excess of colony needs will be accumulated as "surplus" honey in the colony and pollen deposited in cells near the brood nest.

Oregon is environmentally very diverse, ranging in habitat from rain forest to desert, from sea level to glaciated mountain peaks. As one would expect, such geographical diversity produces many different botanical assemblages which vary tremendously in the rewards they offer bees. The most important nectar and pollen producing plant species vary throughout the state and in their attractiveness in different areas. See map 1; Beekeeping Areas of Oregon.

1

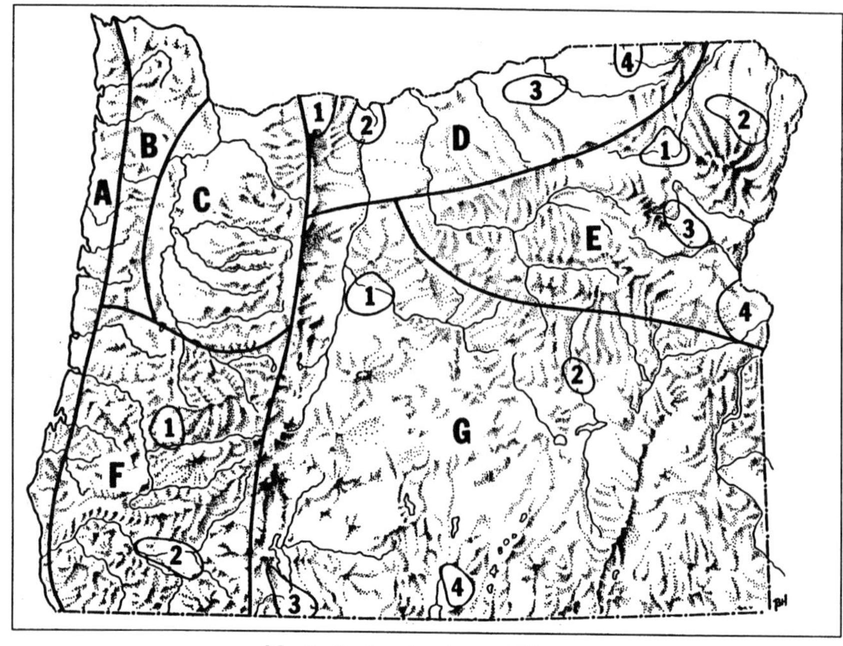

Map 1. Beekeeping Areas of Oregon

A. The coastal area
B. The northern Coast Mountains
C. The Willamette Valley and foothills
D. The Columbia Basin:
    1. Hood River district
    2. The Dalles district
    3. Umatilla-Boardman district
    4. Milton-Freewater district
E. The Blue Mountain area:
    1. The Grande Ronde Valley
    2. The Wallowa Valley
    3. Powder River Valley
    4. Malheur district
F. The southwestern mountains:
    1. The Umpqua Valley
    2. The Rogue River Valley
G. The central desert area:
    1. The Redmond district
    2. The Burns district
    3. The Klamath Basin
    4. The Lakeview district

In 1942 H.A. Scullen and G.A. Vansell wrote the first catalogue of Oregon's bee forage plants, *Nectar and Pollen Plants of Oregon.* Beekeeper cooperation and research by the authors resulted in an overview of the nectar and pollen plants of benefit to honey bees at that time. This classic publication has long been out of print, and Oregon's honey bee environment has changed in the years since 1942, yet the demand persists for a reference of this type. Books and articles dealing with the flora of other states do not address the specific climatic and botanical conditions of Oregon, but may contain background information on some of the plants of the region.

Why do plants produce nectar and pollen?

Success, in the world of flowering plants, is measured in terms of reproductive accomplishment. In most cases pollen must be transferred, either between flowers or within a flower, to a receptive stigma where it germinates and effects pollination of the flower. The resulting seed then develops, ripens, is shed, and germinates into another plant. Nectar fits into the scheme as the lure to entice honey bees and other insects into the flower. In the process of gathering nectar, honey bees transfer pollen between flowers, their useful habit of 'flower fidelity' ensuring the pollen is transferred to the same type of flower as the donor bloom. Honey bees also actively collect pollen for food, during which many pollen grains adhere to the bee's hairy body and are transferred to other stigmas as the bee brushes past.

There is, then, a very complementary relationship between bees and flowers: bees depend on flowers for their nutritional needs, and some flowers rely on bees and other insects to ensure reproductive success.

What is a honey plant?

Honey plants are classified as major, secondary, and minor, according to their contribution to a bee colony in a given region. A major honey plant species may be defined as one which can be expected to yield a surplus of honey in most years. Also in this category should be placed the plant species which reliably contribute to colony buildup in the spring. The number of major honey producing plant species is usually small in a given area or region. Oregon's major honey plants are alfalfa in the east, fireweed in the Cascade and Coast mountains, star thistle in southern Oregon, and seed clovers, vetch and berries in the Willamette Valley.

Agricultural plantings are frequent contributors to a region's surplus honey production. The past 40 years have seen major agricultural changes which have reduced surplus honey production, especially in western Oregon. Legume seed production in the Willamette Valley, while still influencing honey yields, has declined significantly from former years. On the other hand, increased vegetable seed farming in central and eastern Oregon has improved honey production potential in those areas.

A secondary honey plant will contribute to the surplus honey in limited geographical areas or in favorable years. Examples in Oregon include maples in the west, rabbitbrush in the central and eastern regions, madrone and vetch in southern Oregon and snowberry in the west and northeast.

Minor sources are those which are important for colony buildup and maintenance, but rare as contributors to surplus honey production. Common minor plants include dandelion, cascara and many wildflower and weed species. In metropolitan areas some landscaping plants, such as abelia, cotoneaster, English laurel, heather and pyracantha, may constitute important minor forage for the few colonies of the urban beekeeper.

All plant communities grow and change. Many valuable nectar and pollen producing plants are intermediate in the progression between original colonizers and a climax community. Fireweed, for example, lasts only about ten years after a forest burn before being replaced by other plants. Logging or agricultural practices may change the value of an area to bees. When mixed forest is replaced by a conifer monoculture, both diversity and bee forage are lost. Conversely, a fallow field planted to hairy vetch may improve local bee pasture.

Nectar Secretion and Pollen Production

The nature of nectar secretion is not well understood, but apparently there are many elements involved, one of which is the general condition of the plant. The current condition of a perennial plant may depend on such distant events as how well it weathered the previous summer's drought, or how badly it was damaged by late spring frosts, or how much carbohydrate was stored in the roots in fall.

Plant growth and nectar secretion are also influenced by soil type, rainfall, solar exposure, temperature extremes and altitude. Those plants which produce nectar and pollen must be present in sufficient floral density, and have their nectar available and attractive, to be considered a significant food source for a honey bee colony. Also, weather during the bloom period must allow sufficient bee flight in order for nectar and pollen to be collected. If the weather is poor, bees may be unable to gather nectar and pollen even from the most prolific sources. Plants which secrete nectar beyond the reach of the honey bee's tongue, or with low sugar content, or scant volume, are rarely used by bees.

In contrast, conditions may seem ideal for a honey flow from a particular species, yet the bees fail to gather a surplus. Fireweed is an example of the unpredictability of nectar plants. In most years surplus honey production from fireweed is average, a few years poor, and about one year in seven, excellent. At present there is no reliable way of foretelling good or poor honey flows.

When fireweed and alsike clover nectar secretion were studied in Oregon, it was found that the nectar sugar concentration was inversely related to the relative humidity; that is, the drier the air the "sweeter" the nectar. Water in nectar tends to evaporate during the day, particularly from flowers with an open shape like fruit tree blossoms, which increases the sugar concentration of the nectar and thus its attractiveness to foraging bees. Relative humidity affects the evaporation rate of water from the nectar and so influences how attractive the nectar is to bees. Also, some plants secrete nectar with higher sugar concentration than others. Examples of high and low nectar sugars are, big-leaf maple (>50%) and pear (5 to 15%).

Release of pollen from a flower's anthers, or dehiscence, is not constant through the day. Pollen may be available for collection as soon as the flower opens, or may not be released until the air warms to a certain level. The protein content of pollen varies among plants from about 10-30%, yet bees collect pollen without regard to its nutritive value. Wind pollinated plants tend to produce lower quality pollen than that of plants dependent on insects for pollination. Some flowers, such as crocus, hollyhock and poppy, are sources of high quality pollen which is extremely attractive to bees, and frequently several bees may be seen working the same bloom.

Nectar and pollen are not necessarily produced by the same flower. Plants may bear distinct male and female flowers, either on the same plant or on separate plants of the same species. Pollen is only produced by male flowers, but nectaries may be present in staminate (male) or pistillate (female) flowers, and even on some leaves and stems.

Beekeeping population

In common with most states, Oregon's beekeeping community consists of a large number of hobbyists who own a few hives each, several hundred sideliners who maintain individual operations of 25 to 200 colonies, and a small population of commercial operators who own the majority of the colonies. See map 2; Colony Distribution.

Most hobbyists keep their colonies on fixed locations and depend on local flora for their surplus honey production. Naturally occurring trees, shrubs, berry

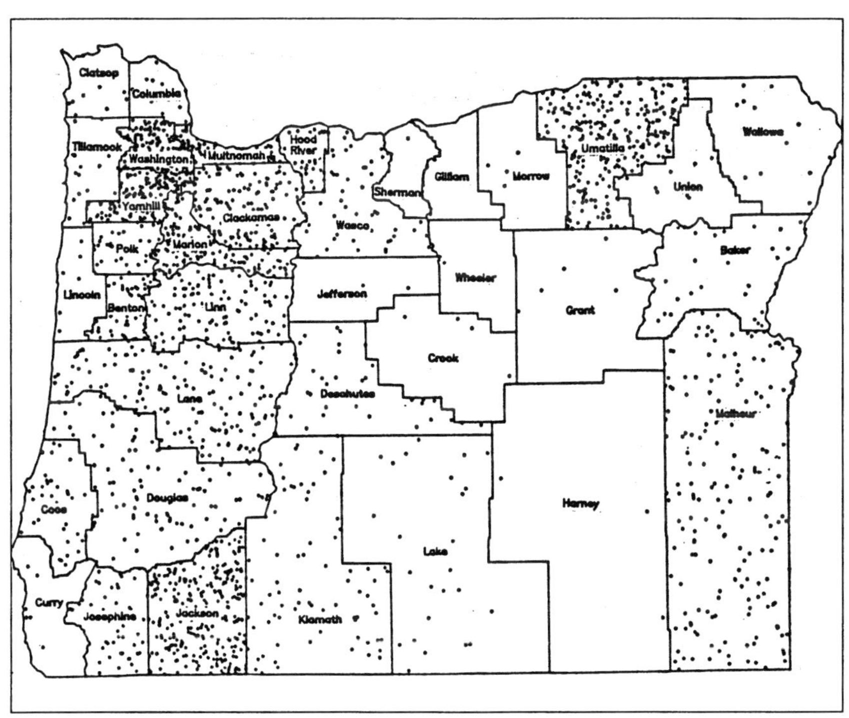

Map 2. Colony Distribution by County, 1988

One dot represents 25 colonies. This map does not indicate colony distribution within the county.

vines and wildflowers are the sustaining forage for a hobbyist's apiary.

Approximately 90% of registered beekeepers in Oregon own fewer than 20 colonies. Many of these colonies are kept in urban or suburban areas, where common landscaping plants may become important nectar and pollen sources for honey bees. While it is not economical to plant solely for bee forage, urban beekeepers may derive much pleasure from watching bees work garden plantings. Particularly valuable are those plants which flower very early or very late in the season, or when little else is in bloom.

Commercial beekeepers rent their honey bee colonies for the pollination of fruit and seed crops, sometimes obtaining a surplus honey crop as a benefit. Most commercial beekeepers in Oregon expect to derive up to two-thirds of their income from pollination rentals and the remainder from honey sales.

While smaller beekeeping operations are found throughout the state, the main centers of commercial beekeeping are concentrated in eastern and southwestern Oregon and the Willamette Valley. Not all locales in the state are favorable for year-round beekeeping. Limiting conditions include seasonal lack of nectar and pollen plants, low temperatures, high altitudes, frequent fog or lack of summer moisture.

Geography

Oregon's Cascade Mountains are an important barrier to maritime and continental air flows, essentially dividing the state into a wetter west and a drier east. West of the Cascades there are climate variations related to the coastal mountains and the latitude. Interior valleys, in the rainshadow of the coastal mountains, are generally drier.

Oregon's soils are influenced by three primary geological sources from which the soils were formed by the action of climate on the geological base over time. Sandstone dominates in the Coast Range, basalts in the Cascades and eastern Oregon plateau, and alluvials in the Willamette Valley and eastern Oregon lake beds.

The types of vegetation or crops supported by the climate and soils differ in the various beekeeping regions of the state. East of the Cascades, hay and seed crops are grown in those irrigated areas where fertile valleys penetrate the high desert plateau. Yellow and white sweet clovers and alfalfa are important honey bee forage species in these areas. Alfalfa seed producing regions in eastern Oregon account for the most significant source of surplus honey and numerous colonies are moved to this area from other parts of the state to take advantage of the summer nectar sources. Rabbit brush, star thistle, rape seed and buckwheat may be locally important supporting bee pasture.

The beekeeping season in southwest Oregon is generally earlier than in the rest of the state. Abundant manzanita contributes nectar and pollen for early spring buildup, and clover, madrone, Himalayan blackberry, vetch and alfalfa are common. Lack of summer moisture and high temperatures limit later pasturage in this region, and many beekeepers move their colonies east of the Cascades or to fireweed locations in the summer. Star thistle can be locally important for surplus honey.

The Willamette Valley is the most agriculturally diverse region of the state. The wide variety of crops grown benefit from honey bee pollination and contribute to overall bee forage. Of principal importance are fruit bloom, vetch and crimson clover grown for seed, white and red seed clover, and berries. Commercially grown acreages of fruits and berries require honey bee pollination for which growers rent honey bee colonies. Clover and vetch seed crops are known to frequently produce surplus honey for the colonies placed in these fields at minimal or no rental fee. The primary honey flow period is usually over by the early part of July when most commercial and sideliner beekeepers move their colonies to the mountains for the fireweed flow, or to eastern Oregon for alfalfa. Blackberries and native wildflowers and shrub species are major honey sources in the Willamette Valley especially for fixed location apiaries.

The Illustrated Dictionary

In revising *Nectar and Pollen Plants of Oregon*, much of the original material has been expanded into an illustrated dictionary of plants used by bees. Listed alphabetically by the plant's most frequently used common name, each entry records the currently correct botanical name and the family to which the plant belongs. Further details on growth-type, bloom period, flower color and distribution will aid in identification of a species' beekeeping importance. Collection of nectar (n), pollen (p), or no nectar (or pollen) (-) is recorded. A question mark (?) means information is lacking or conflicting. Other related articles of interest conclude each entry.

Each plant entry is recorded in this format:

Common name
Botanical name
FAMILY TO WHICH THE PLANT BELONGS
Growth type of the plant, e.g. annual
Bloom period   Flower color
Nectar and/or pollen production
Distribution of the plant
Additional remarks

This work is not intended as a botanical key. For precise identification and more information, the reader is referred to the books recommended in the Bibliography, from which many of the illustrations in this guide have been drawn. Botanical details for the listed plants may be obtained from reference works recorded in the Illustration Credits at the beginning of the book.

Fig. 1 *Abelia* x *grandiflora*

Abelia, Glossy
Abelia x *grandiflora* (Andre) Rehd.
CAPRIFOLIACEAE
Shrub
June to October
Pink, white
n p

As planted

Common hedge or shrub. Also used in highway landscaping.

Fig. 2 *Alnus rubra*; a,leaf; b, old pistillate catkins; c, staminate catkins.

Alder   Tree
Alnus spp.   Early spring
BETULACEAE   Green catkins   - p

Mostly in wet areas

General in western Oregon but limited to mountain stream banks in eastern Oregon.

Alfalfa
Medicago sativa L.
LEGUMINOSAE
Perennial
Summer
Blue, purple
n p

Mainly Central & Eastern & Klamath areas

Grown as seed and hay crop. A source of water white, fine quality honey. Little pollen collected. Nectar sugar concentration 40-45%. Unreliable producer at high elevations.

Fig. 3 *Medicago sativa*

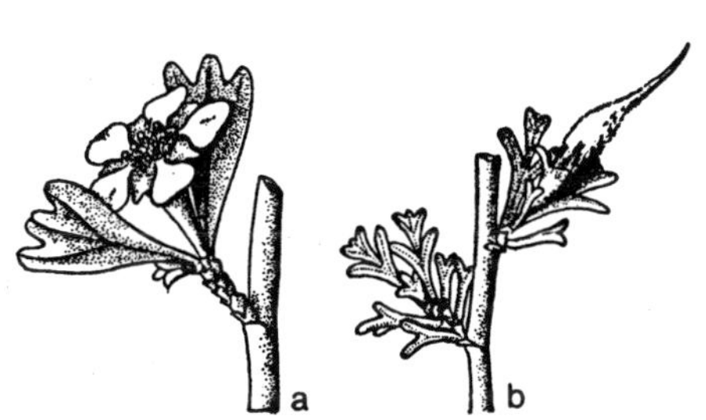

Antelope Brush
Purshia tridentata (Pursh) DC.
ROSACEAE
Low shrub
Late April
Yellow
n p

Fig. 4 *Purshia tridentata*; a, flowering stem;
b, fruiting branchlet.

Eastern Oregon

Sandy, arid ground. Bloom begins between late April and mid May, depending on elevation. Also called bitter brush and buck brush. Nectar sugar concentration from 30 to 50%.

Fig. 5 *Malus sylvestris*; a, blossom; b, fruit.

Apple   Tree
Malus sylvestris Mill.   April and May
ROSACEAE   White, pink, red   n p

Commercially planted in Rogue River Valley, Willamette Valley and Columbia Gorge.

Approximately 10,000 acres grown commercially in Oregon. Most commercial varieties are self-sterile and require insect pollination. Nectar sugars from 45 to 56%.

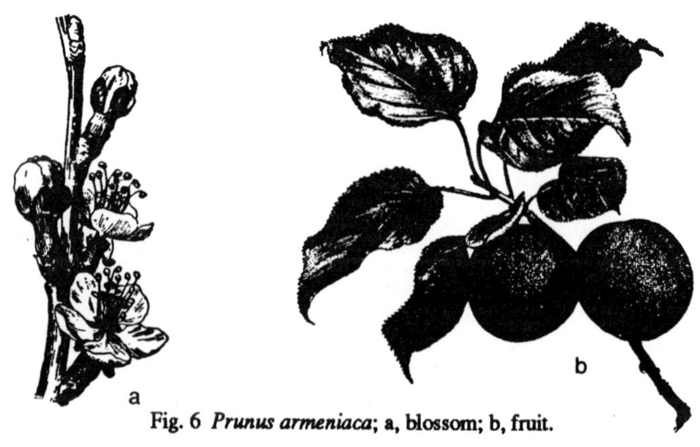

Fig. 6 *Prunus armeniaca*; a, blossom; b, fruit.

Apricot   Tree
Prunus armeniaca L.   Early spring
ROSACEAE   White   n p

As planted

Only a few hundred commercial acres in the state.

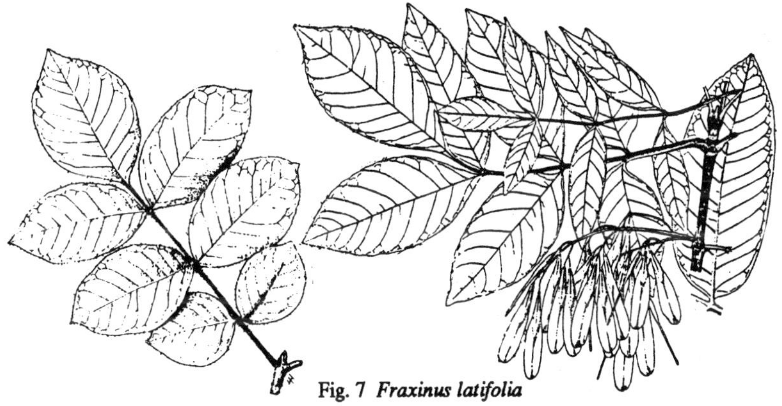

Fig. 7 *Fraxinus latifolia*

Ash, Oregon
Fraxinus latifolia Benth.
OLEACEAE

Tree
Spring
Greenish yellow   - p

Western Oregon

Confined largely to low flooded areas. A liberal quantity of pollen is produced by staminate flowers.

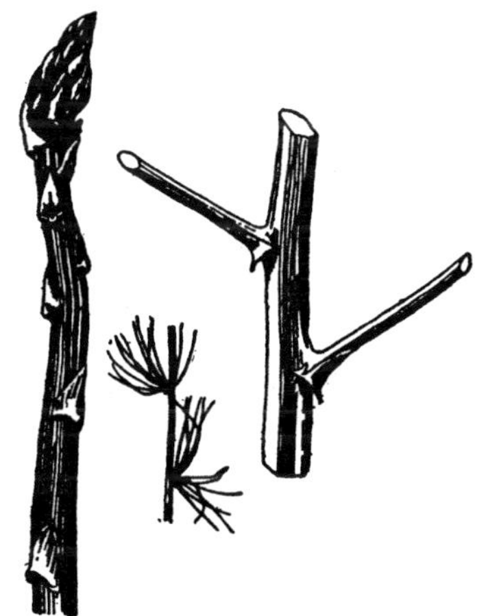

Asparagus
Asparagus officinalis L.
LILIACEAE
Perennial herb
Summer
Yellow
n p

Fig. 8 *Asparagus officinalis*

Willamette, Hood River valleys, Wasco and Umatilla counties.

An estimated 1,600 commercial acres in 1985. Honey is light in color and reported to be of fair quality.

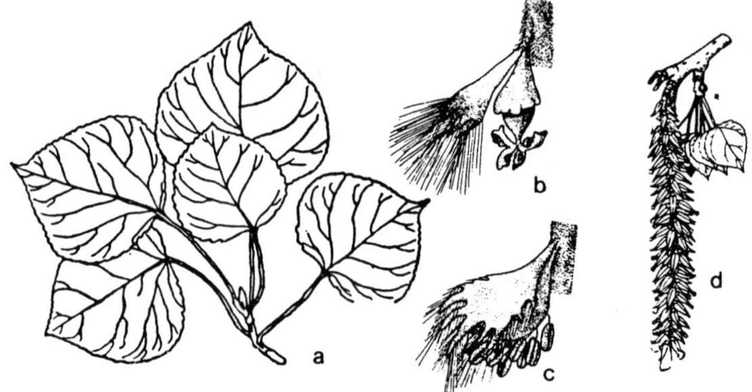

Fig. 9 *Populus tremuloides*; a, leaf; b, pistillate flower; c, staminate flower; d, catkin.

Aspen, Quaking
Populus tremuloides Michx.
SALICAEAE

Tree
Early spring
Greenish yellow catkins   - p

Widespread east of the Cascades. Occasional in the west

Several cultivated species may be locally important. Male and female flowers appear before the plant leafs out and are on separate trees.

Aster
Aster spp.
COMPOSITAE
Perennial herb
Late summer
White, reds, blues
n p

State wide with several cultivated species locally important.

Late source of nectar and yellow pollen.

Fig. 10 *Aster subspicatus*; a, flowering stem; b, basal leaf.

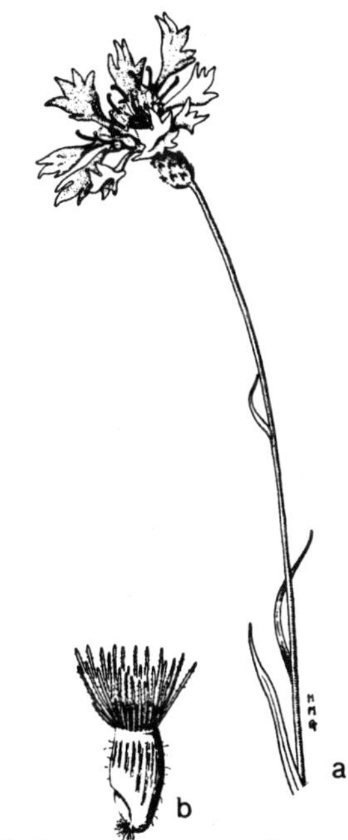

Fig. 11 *Centaurea cyanus*; a, flower; b, fruit.

Bachelor's Button
Centaurea cyanus L.
COMPOSITAE
Annual
Summer
Blue, white, reds
n p

Cultivated. Escaped and now naturalized especially west of the Cascades

The honey is yellowish-green and has a strong flavor. Nectar sugars recorded up to 53%. Cornflower is another frequent common name for this species.

Fig. 12 *Balsamorhiza deltoidea*

Balsam Root
Balsamorhiza deltoidea Nutt.
COMPOSITAE
Perennial herb
Spring
Yellow
n p

Widespread. Open hillsides in general

Looks like a dwarf sunflower. Nectar sugars as high as 46%. Six other species of *Balsamorhiza* are recorded from Oregon.

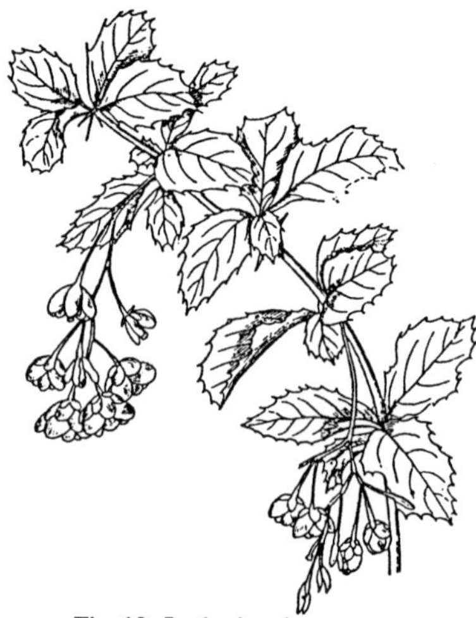

Barberry
Berberis spp.
BERBERIDACEAE
Shrub
Spring
Yellow, orange
n p

Fig. 13 *Berberis aristata*

As planted

Several introduced species used as ornamentals. Worked freely by bees but not common enough to be of significant value. See Oregon grape.

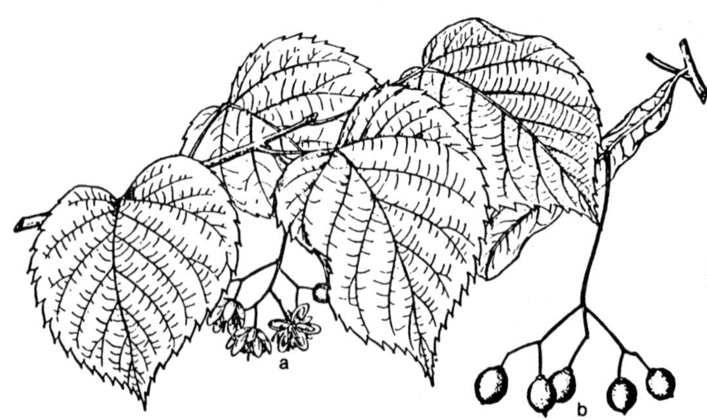

Fig. 14 *Tilia americana*; a, flowers; b, fruit.

Basswood
Tilia spp.
TILIACEAE

Tree
Early summer
Cream
n p

As planted

Both the American species (*T. americana* L.) and the European species (*T. europaea* L.) are used as shade trees. Often called linden or lime tree. Nectar sugars range around 35%.

Fig. 15 *Phaseolus* spp.

Bean
Phaseolus spp.
LEGUMINOSAE
Annual
Summer
White, red
n ?

As planted

Many different bean species are commercially grown. Only the lima bean has any significant attraction for honey bees and it is grown in limited amounts.

Fig. 16 *Bidens frondosa*

Beggar-ticks
Bidens spp.
COMPOSITAE
Annual or perennial herb
July to October
Pale yellow
n p

Damp ground throughout the state

Seldom common enough to be of much value, although parts of Union County may be an exception. Nectar sugar concentration 40%. Also called Stick-tight.

Bindweed, Black
Polygonum convolvulus L.
POLYGONACEAE
Annual
May to October
Pink
? ?

A widespread weed

An introduced weed of European origin. Reported as a being used by honey bee foragers in the east. Not known to be worked by bees in the Pacific Northwest.

Fig. 17 *Polygonum convolvulus*

Birch, Bog
Betula glandulosa Michx.
BETULACEAE
Shrub
Spring to early summer
Green catkins
- p

Fig. 18 *Betula glandulosa*; a, leaf; b, flowering
branchlet.

Stream banks, bogs and marshes

Also known as scrub birch. Habitat out of range of most honey bees.

16

Blackberry, Evergreen
Rubus laciniatus Willd.
ROSACEAE
Shrub
May through June
White, pale pink
n p

Western Oregon and occasionally east of
Cascades

Grown commercially and widely escaped.
Nectar sugars about 36%. A major source
of surplus honey in Western Oregon.

Fig. 19 *Rubus laciniatus*; a, flowering
stem; b, lower leaf.

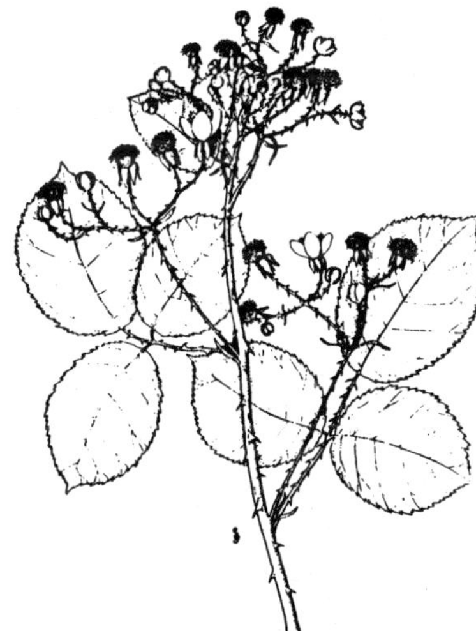

Blackberry, Himalaya
Rubus procerus P.J. Muell.
ROSACEAE
Vigorous shrub
Late May through June
White
n p

Fig. 20 *Rubus procerus*

Western Oregon

Grown commercially and widely escaped. A major source of surplus honey in
the Willamette Valley, Rogue Valley and coast. Nectar sugars measured at
27%.

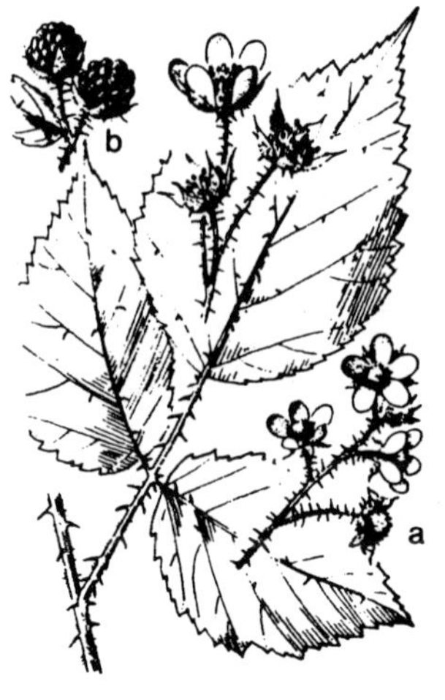

Blackberry, Native
Rubus ursinus C.& S.
ROSACEAE
Trailing shrub
Late April and May
White

n p

Western Oregon

Also known as Pacific blackberry or dewberry. Oregon's only native blackberry. Excellent nectar source. Nectar sugars range between 40 and 60%.

Fig. 21 *Rubus ursinus*; a, flowers; b, fruit.

Bluebell, Wild
Campanula scouleri Hook.
CAMPANULACEAE
Perennial herb
June to August
Blue

n p

Fig. 22 *Campanula scouleri*

Mostly west of Cascade summit

Has been reported as an occasional honey plant in north-western Oregon, but is probably of little value in the Pacific Northwest.

Blue Curls
Trichostema lanceolatum Benth.
LABIATAE
Annual
Summer
Blue
n p

Rogue River and Willamette valleys

Important producer in California but of little value in the Pacific Northwest. Also known as vinegar weed and camphor weed.

Fig. 23 *Trichostema lanceolatum*

Borage, Common
Borago officinalis L.
BORAGINACEAE
Annual
June to frost
Blue
n p

Fig. 24 *Borago officinalis*

As planted

A European honey plant introduced as a garden species and now escaped. Of minor value. Very attractive to honey bees.

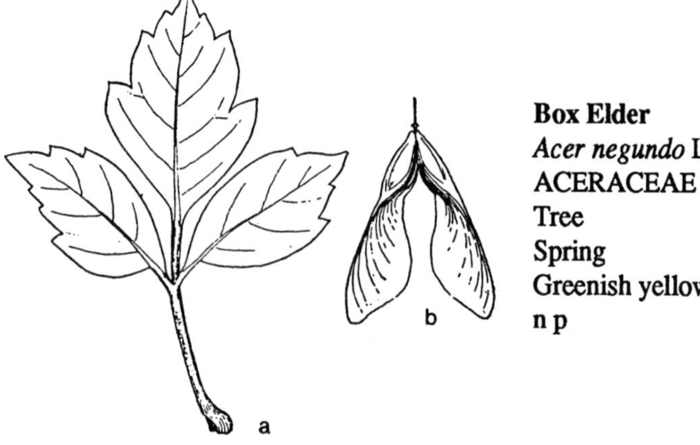

Fig. 25 *Acer negundo*; a, leaf; b, fruit.

As planted. More common in eastern Oregon. Occasionally escaped and naturalized.

Not a heavy nectar producer. Only the male trees provide pollen.

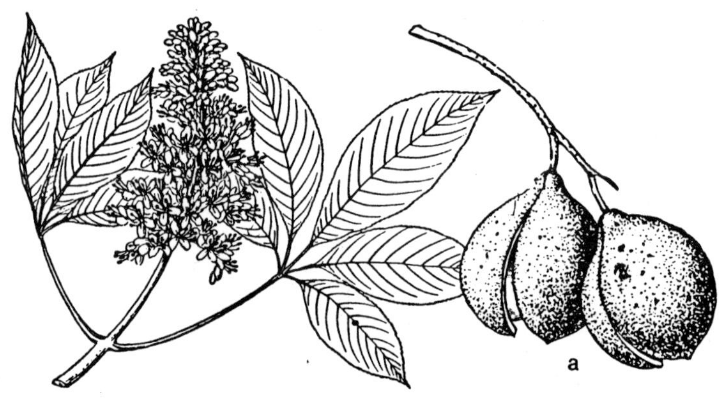

Fig. 26 *Aesculus californica*; a, fruit.

Buckeye, California Tree
Aesculus californica (Spach) Nutt. Early summer
HIPPOCASTANACEAE Cream n p

As occasionally planted. Rare in the Pacific Northwest.

Very destructive to honey bees in California. Not native to Oregon. The horse chestnut, related to buckeye and used sparingly as a shade tree in Oregon, is not toxic to bees.

Buckwheat
Fagopyrum esculentum Moench
POLYGONACEAE
Annual
Summer
White
n p

As occasionally planted. A surplus honey producer where grown.

A native of Asia and grown commercially for seed. Furnishes very little pollen but copious dark nectar. Nectar sugars about 50%.

Fig. 27 *Fagopyrum esculentum*

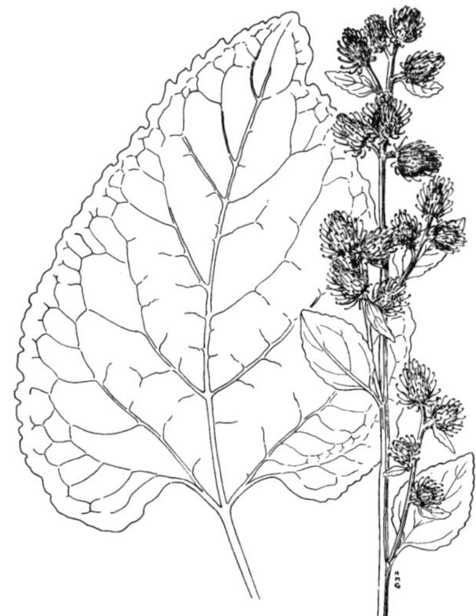

Burdock, Common
Arctium minus (Hill) Bernh.
COMPOSITAE
Biennial
Summer
Purple
n p

Fig. 28 *Arctium minus*

General throughout state, but very limited in density.

Attractive to honey bees but so infrequent to be of little value.   An introduced weed species of Eurasian origin.

Burnet
Sanguisorba spp.
ROSACEAE
Annual herb
Summer
Green, white, purple
- p

Most common in the Willamette Valley but generally widespread.

Considerable pollen is collected from the various species. Used in conservation plantings.

Fig. 29 *Sanguisorba occidentalis*;
a, flowering stem; b, flower.

Buttercup
Ranunculus spp.
RANUNCULACEAE
Annual or perennial herbs
March and April
Yellow
n p

Fig. 30 *Ranunculus occidentalis*

General

Of some value to honey bees because of the early bloom. Furnishes very little pollen.

Cabbage
Brassica oleracea L.
CRUCIFERAE
Herb
Late spring to summer
Yellow
n p

Fig. 31 *Brassica oleracea*

As planted commercially for seed. Western and central Oregon.

An excellent honey plant with several thousands of acres devoted to commercial seed production. Honey granulates readily.

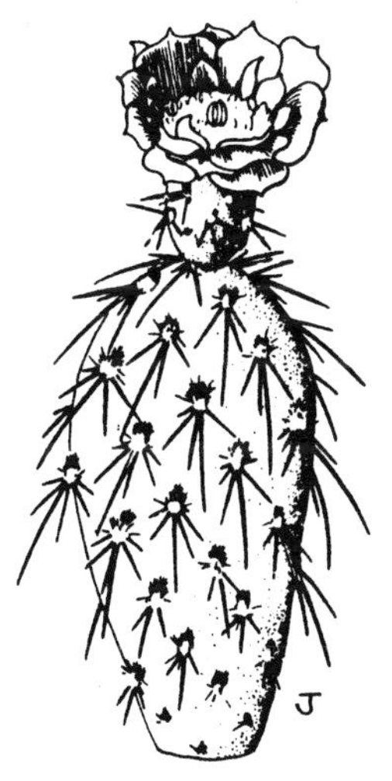

Cactus, Prickly-pear
Opuntia spp.
CACTACEAE
Spiny perennial
May through June
Yellow, white
n p

Eastern Oregon deserts

Two species of cacti in Oregon are commonly called prickly-pear. They are probably of some value as pollen sources where common.

Fig. 32 *Opuntia polyacantha*

Fig. 33 *Daucus carota*

Carrot
Daucus carota L.
UMBELLIFERAE
Biennial
Late June to mid August
White
n p

Grown commercially for seed in central and eastern Oregon

Several thousands of acres grown for seed require honey bee pollination. Produces a strong flavored honey used as colony winter feed or for the bakery trade. Wild carrot, worked mainly for pollen, is also called Queen Anne's Lace.

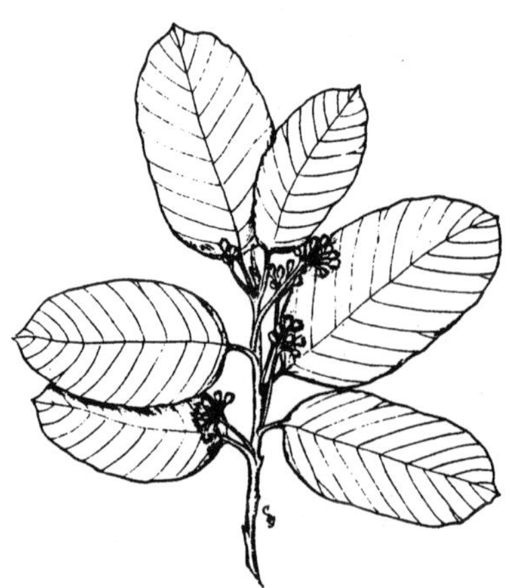

Fig. 34 *Rhamnus purshiana*

Cascara
Rhamnus purshiana DC.
RHAMNACEAE
Small tree
April to early June
Green
n p

Common in western Oregon, uncommon east of Cascades

Occasional surplus honey crops reported. Honey is light amber in color. Bloom may last up to six weeks. Commonly called chittim.

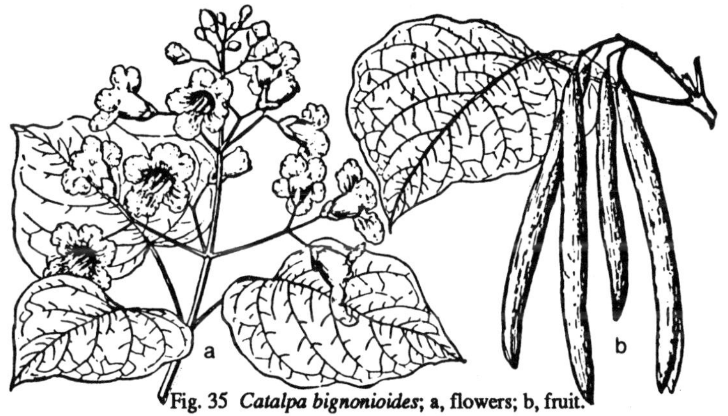

Fig. 35 *Catalpa bignonioides*; a, flowers; b, fruit.

Catalpa   Tree
Catalpa spp.   June
BIGNONIACEAE   White   n p

As planted as an ornamental

Too few trees are planted to be of much value, but the tree is a heavy yielder of nectar.

Catnip
Nepeta cataria L.
LABIATAE
Perennial herb
Summer
White, lavender
n -

Widely cultivated throughout the state

A heavy producer of nectar, but not common enough to be of much value. Nectar sugars reported at 29%. Rarely utilized as a pollen source.

Fig. 36 *Nepeta cataria*; a, flower.

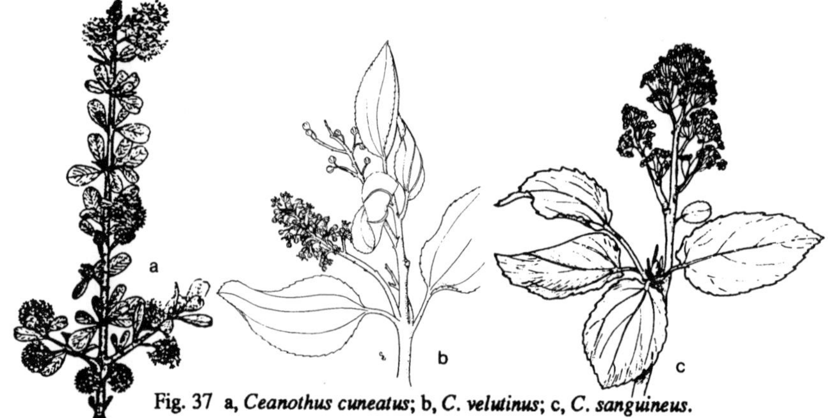

Fig. 37 a, *Ceanothus cuneatus*; b, *C. velutinus*; c, *C. sanguineus*.

Ceanothus   Shrub
Ceanothus spp.   Early May
RHAMNACEAE   White, pink   n p

Common mountain and foothill species.

Three common species are major contributors to spring buildup; *C. cuneatus* (greasewood) pollen only, *C. velutinus* (cinnamon bush), and *C. sanguineus* (buckbrush).

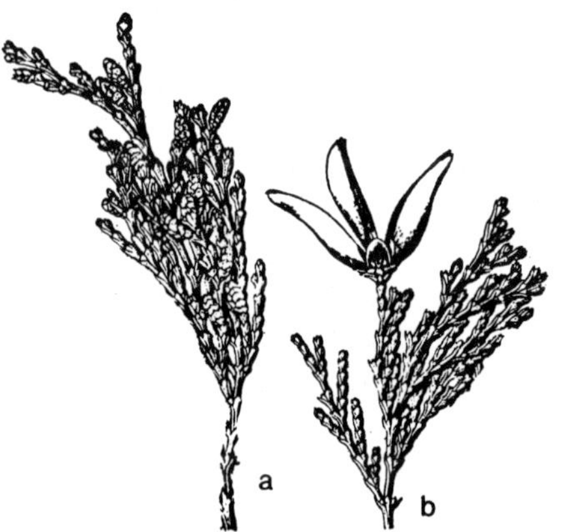

Cedar, Incense
Calocedrus decurrens (Torr.) Florin
CUPRESSACEAE
Tree
Winter & early spring
Green catkins

- -

Fig. 38 *Calocedrus decurrens*; a, male flowers;
     b, fruiting branch.

Primarily in the Cascades

No floral nectar or pollen. However, honey bees have been known to collect honeydew from the incense cedar scale (*Xylococculus macrocarpae* Coleman) in the Rogue River Valley.

Celery
Apium graveolens L.
UMBELLIFERAE
Biennial
Summer
Green
n -

As commercially grown

A good producer of nectar when grown for seed. However, very little seed is grown in the Pacific Northwest.

Fig. 39 *Apium graveolens*

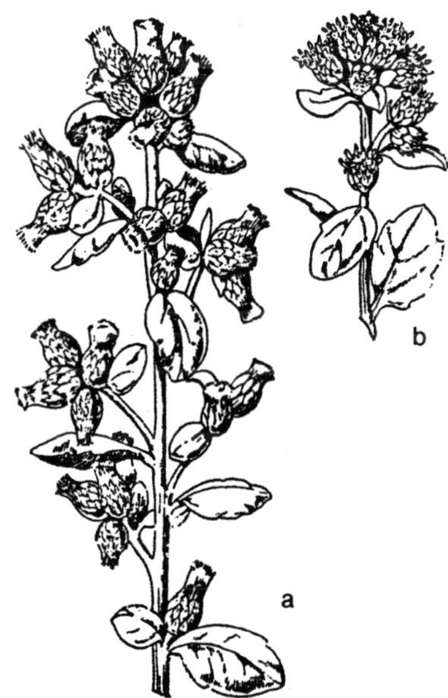

Chaparral Broom
Baccharis pilularis DC.
COMPOSITAE
Shrub
August to September
White, yellow
n p

Along the coast and rarely inland

Also called wine bush and coyote brush. Yields nectar and pollen in parts of California, but its value in Oregon has not been confirmed.

Fig. 40 *Baccharis pilularis*; a, pistillate branchlet; b, staminate branchlet.

Fig. 41 *Prunus emarginata* var. *erecta*

Cherry, Bitter
Prunus emarginata (Dougl.)
Walp.
ROSACEAE
Tree or large shrub
Spring
White
n -

Widespread in open woods.

Of minor importance to honey bees due to limited abundance. Nectar sugars reported between 40 to 50%. Also known as wild cherry.

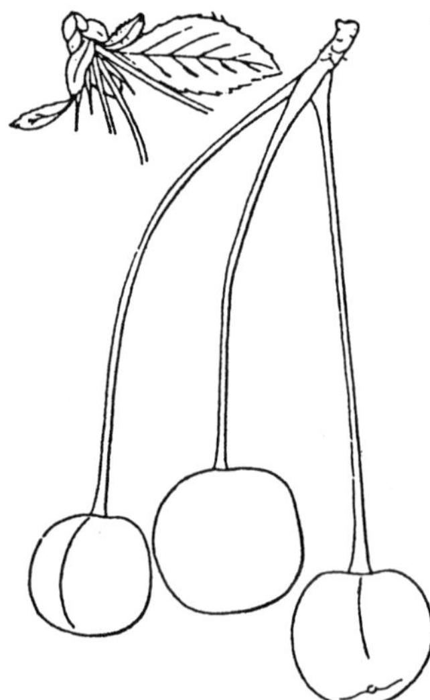

Fig. 42 *Prunus cerasus*

Cherry, Sour
Prunus cerasus L.
ROSACEAE
Tree
April
White
n p

As cultivated

These cherries bloom about one week later than sweet cherries and are also attractive to honey bees. Nectar sugar concentration 15-43%. Also called pie or tart cherries.

Cherry, Sweet
Prunus avium L.
ROSACEAE
Tree
April
White
n p

Fig. 43 *Prunus avium*

Cultivated. Willamette Valley, Wasco County.

12,000 acres of commercial sweet cherries are grown in the state. Honey bee colonies rented for cross pollination as most varieties are self-sterile. Nectar sugars range from 15 to 60%.

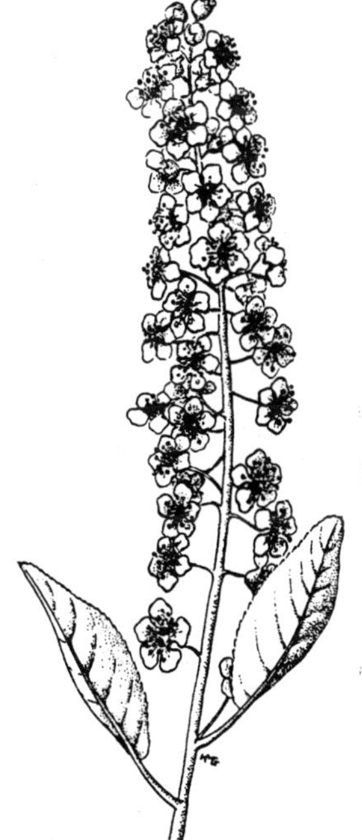

Cherry, Western Choke
Prunus virginiana L. var. *demissa* (Nutt.) Torr.
ROSACEAE
Tree or shrub
Spring
White
n ?

Widespread west of the Cascades

Only occasionally worked by honey bees. May be an important minor plant in Southern Oregon.

Fig. 44 *Prunus virginiana*

Chestnut, Horse
Aesculus hippocastanum L.
HIPPOCASTANACEAE
Tree
Mid-May to mid-June
Cream, reds
n p

As planted as an ornamental

Nectar flow lasts about four weeks. Nectar sugar concentration varies from 30% to 75%. Abundant brick-red pollen produced. The American Chestnut, *Castanea dentata*, of the family Fagaceae, is now rare because of disease.

Fig. 45 *Aesculus hippocastanum*

Chickweed, Common
Stellaria media (L.) Cyr.
CARYOPHYLLACEAE
Annual
February to October
White
n p

General in gardens and as an escaped species.

An introduced weed species of minor importance, but useful early and late in the season. Nectar sugars reported at 50%.

Fig. 46 *Stellaria media*

Chicory
Cichorium intybus L.
COMPOSITAE
Perennial herb
July to October
Blue
n p

Widespread

An introduced weed species of Eurasian origin. Important minor plant in Rogue Valley because of drought tolerance.

Fig. 47 *Cichorium intybus*; a, flowering stem; b, basal leaf.

Fig. 48 *Castanopsis chrysophylla*; a, flowering branch; b, fruit.

Chinquapin, Giant   Shrub or tree
Castanopsis chrysophylla (Dougl.) A.DC. April to June
FAGACEAE   Cream   n p

More common in western Oregon.

A source of spring pollen and nectar for colony growth. Very attractive to honey bees. May bloom through summer.

Clematis
Clematis ligusticifolia Nutt.
RANUNCULACEAE
Woody vine
Early summer
White
n p

Fig. 49 *Clematis ligusticifolia*; a, seed.

General but more common in eastern Oregon canyons.

Also known as Traveler's Joy or White Virgin's bower. Of limited value for bee forage.

Cleome, Yellow
Cleome lutea Hook.
CAPPARIDACEAE
Annual
May to frost
Yellow
n p

East of the Cascade Ridge.

Grows commonly in subirrigated low ground. More common below 2,000 feet elevation. Nectar sugars are low. A dark, strongly flavored honey. See also Rocky Mountain bee plant.

Fig. 50 *Cleome lutea*

Clover, Arrowleaf
Trifolium vesiculosum Savi
LEGUMINOSAE
Annual
July and August
White, pink
n p

As planted

Some varieties produce a good crop of white honey. Honey bees are essential for pollination.

Fig. 51 *Trifolium vesiculosum*

Clover, Bur
Medicago hispida Gaertn.
LEGUMINOSAE
Annual herb
March through June
Yellow
n -

Fig. 52 *Medicago hispida*

General especially west of the Cascades

Appears to be of little value to honey bees in the Pacific Northwest. A native of Europe.

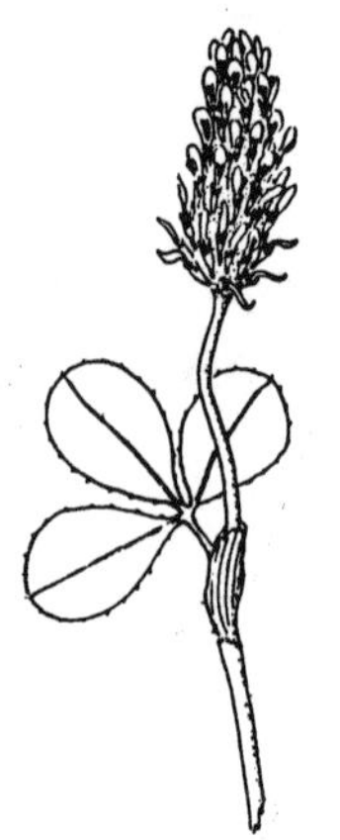

Clover, Crimson
Trifolium incarnatum L.
LEGUMINOSAE
Annual
May
Crimson
n p

Fig. 53 *Trifolium incarnatum*

Grown commercially in the Willamette Valley

The first of the seed clovers to bloom. Occasionally produces a major honey surplus. Always of value as a building flow and very popular with beekeepers in the northern half of the Willamette Valley.

Clover, Red
Trifolium pratense L.
LEGUMINOSAE
Perennial herb
May to August
Red
n p

General

Grown for hay and seed in Oregon. Honey bees gather much dark brown pollen from the self-sterile flowers, and some new varieties produce a moderate nectar flow.

Fig. 54 *Trifolium pratense*

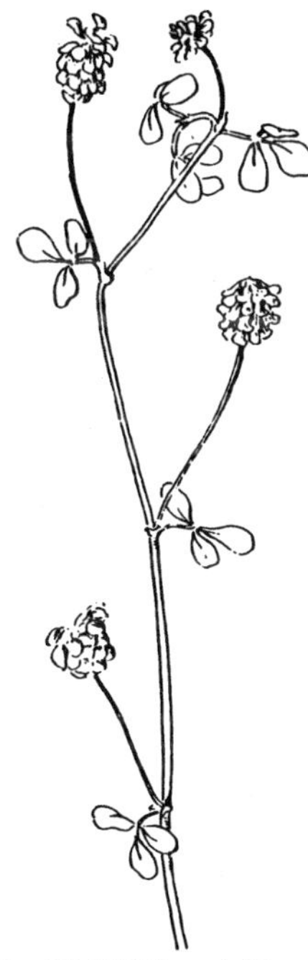

Clover, Small Hop
Trifolium dubium Sibth.
LEGUMINOSAE
Annual
April to September
Yellow
- -

Common in lawns, roadsides and waste areas

Seldom visited by honey bees. A European introduction.

Fig. 55 *Trifolium dubium*

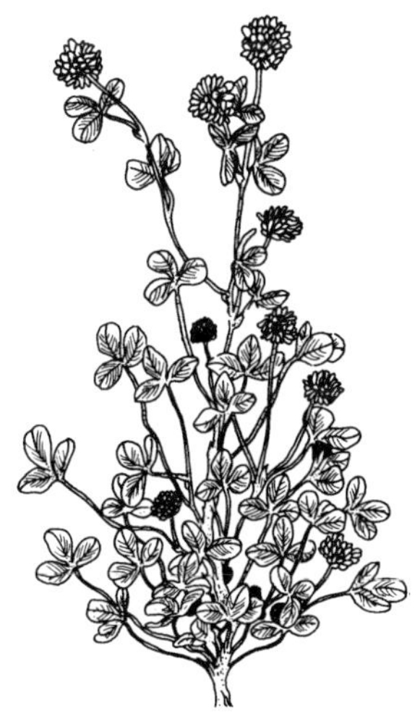

Clover, Strawberry
Trifolium fragiferum L.
LEGUMINOSAE
Perennial
April to July
Purple
n p

Chiefly in eastern Oregon

Bees work the plant freely for nectar (33% sugars) and greenish-brown pollen. Planted as pasture on alkali soil in eastern Oregon and frequently occurs as a volunteer in moist ditches.

Fig. 56 *Trifolium fragiferum*

Clover, Three Toothed
Trifolium tridentatum Lindl.
LEGUMINOSAE
Annual
April to July
Purple
n ?

Western Oregon

Of minor importance.

Fig. 57 *Trifolium tridentatum*

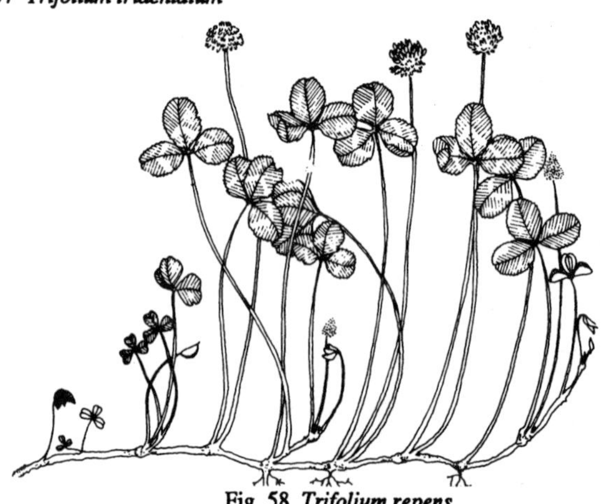

Fig. 58 *Trifolium repens*

Clover, White   Perennial
Trifolium repens L.   May to September
LEGUMINOSAE   White   n p

As planted and naturalized

A commercial seed crop in the Willamette Valley and a common lawn and pasture volunteer. Honey surpluses are not commonly produced. Nectar sugar concentration about 45%.

 Fig. 59 *Xanthium spinosum*

Cocklebur
Xanthium spp.
COMPOSITAE
Annual
April to October
Green
? ?

Northern Oregon

Two species occur in Oregon but have not been recorded as of value. The spiny cocklebur (*X. spinosum* L.) yields nectar in California.

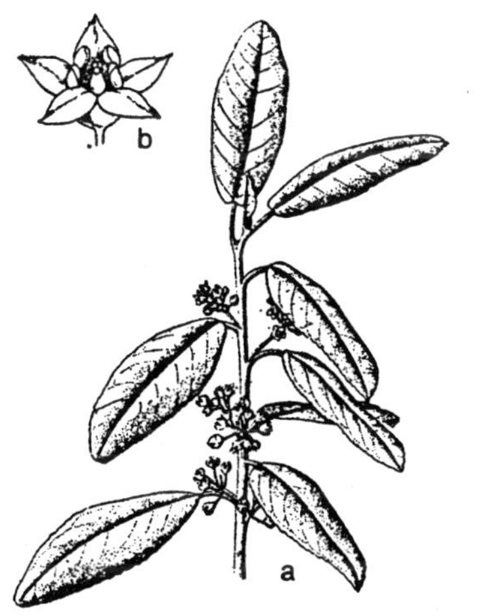 Fig. 60 *Rhamnus californica*; a, flowering branch; b, flower.

Coffee Berry
Rhamnus californica Esch.
RHAMNACEAE
Shrub
May to July
Green
n p

Rogue River Valley

Reported as a very minor source in southwestern Oregon.

Fig. 61 *Rudbeckia occidentalis*

Cone Flower
Rudbeckia occidentalis Nutt.
COMPOSITAE
Tall perennial herb
June to August
Dark purple
n p

General in low mountains
Of minor importance.

Fig. 62 *Zea mays*

Corn
Zea mays L.
GRAMINEAE
Annual
Midsummer
Cream anthers
- p

As planted commercially. Willamette Valley, eastern Oregon.

A pollen source only. Pollen foraging is done in the mornings.

Fig. 63 *Cotoneaster lacteus*; Parney cotoneaster.

Cotoneaster
Cotoneaster spp.
ROSACEAE
Woody shrub
Spring and summer
White, pink
n p

As planted

Over 20 species common in landscaping use in Oregon. Ground covers and shrubs are very attractive to bees.

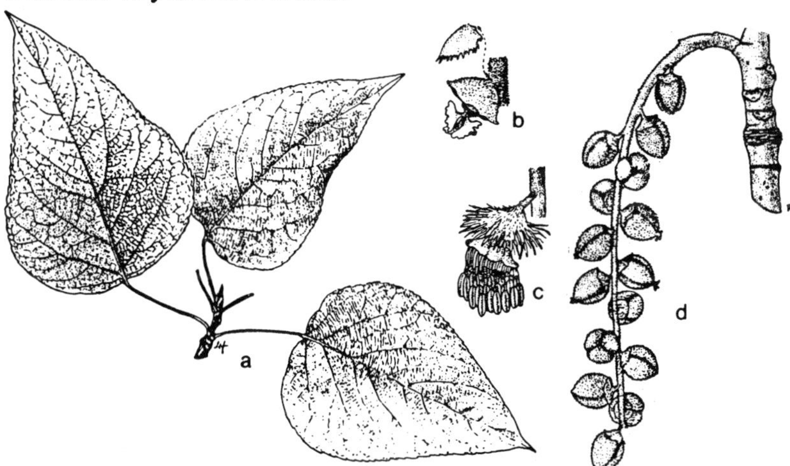

Fig. 64 *Populus trichocarpa*; a, leaves; b, pistillate flower; c, staminate flower; d, catkin with seed pods.

Cottonwood, Black   Tree
Populus trichocarpa Torr. and Gray   Early spring
SALICACEAE   Green catkins   - p

River bottoms and streamsides throughout the state

An important source of aromatic propolis.

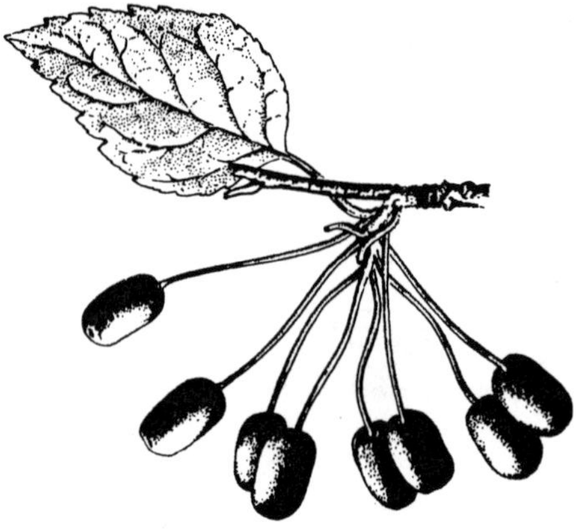

Crabapple, Wild
Pyrus fusca Raf.
ROSACEAE
Tree
Spring
White
n p

Fig. 65 *Pyrus fusca*; fruiting branch.

West of Cascades

Not common enough to be of much value.

Crocus
Crocus spp.
IRIDACEAE
Perennial herb
Early spring or autumn
White, blues, yellows
n p

Fig. 66 *Crocus sativus*

As planted

Very attractive to honey bees, but far too limited to be of much value.

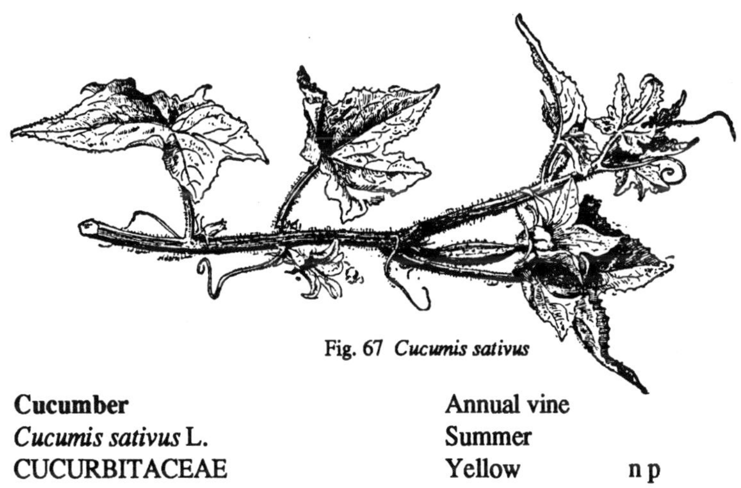

Fig. 67 *Cucumis sativus*

Cucumber   Annual vine
Cucumis sativus L.   Summer
CUCURBITACEAE   Yellow   n p

As planted commercially

A monoecious species (male and female flowers separated on the same plant).
Benefits greatly from honey bee pollinating activities. A poor yielder of either
nectar or pollen.

Currant, Red Flowering
Ribes sanguineum Pursh
SAXIFRAGACEAE
Shrub
Late March to early May
Red
n p

Fig. 68 *Ribes sanguineum*

West of the Cascades

Not very attractive to bees. Nectar sugar measured at 19%. A most popular
forage for hummingbirds.

Fig. 69 *Aster novae-angliae*

Daisy, Michaelmas
Aster spp.
COMPOSITAE
Perennial
Fall
Blues, reds, white
n p

As planted

Common in gardens but not abundant
enough to be of much value. Very
attractive to bees.

Fig. 70 *Taraxacum officinale*

Dandelion
Taraxacum officinale Weber
COMPOSITAE
Biennial herb
April to late summer
Yellow
n p

Throughout the state where annual
moisture is not a limiting factor.
Cosmopolitan weed from Eurasia. A
heavy yielder of nectar and bright orange
pollen. Nectar sugars range between 45
and 50%.

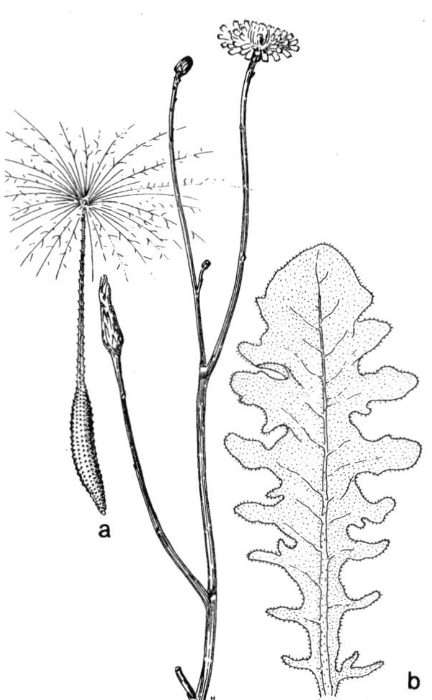

Fig. 71 *Hypochaeris radicata*; a, fruit;
b, leaf.

Dandelion, False
Hypochaeris radicata L.
COMPOSITAE
Perennial herb
Summer into autumn
Yellow
n p

Chiefly in western Oregon

Also known as spotted cat's ear. Abundant in lawns. Light amber nectar. It should not be confused with the common dandelion. Nectar sugars range between 35 and 50%.

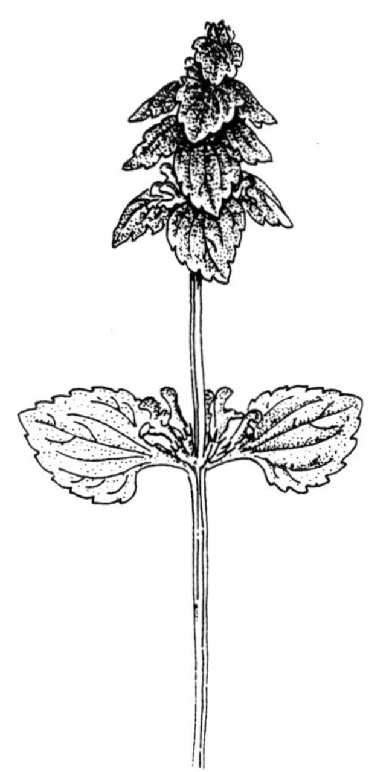

Fig. 72 *Lamium purpureum*

Dead-nettle, Red
Lamium purpureum L.
LABIATAE
Annual
Early spring
Purple
n p

Locally abundant in moist places

Common in cultivated areas and orchards. Bright red-orange pollen.

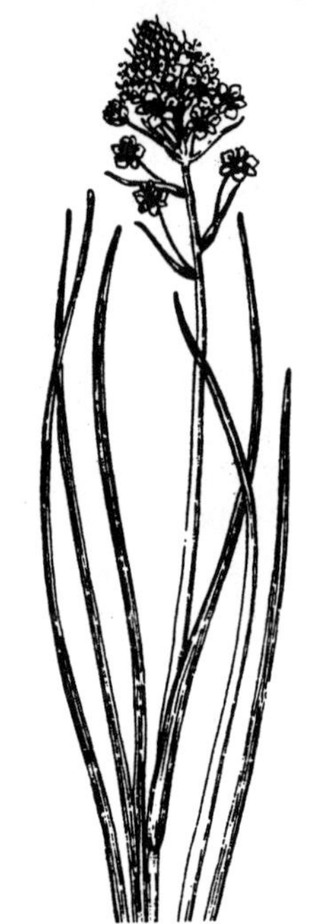

Fig. 73 *Zygadenus venenosus*

Death Camas
Zygadenus venenosus Wats.
LILIACEAE
Perennial herb
March to July
Cream
? ?

Most areas of the state in moist areas.

This plant has been reported as toxic to foraging bees, but no evidence for this claim exists. Not very attractive to bees.

Fig. 74 *Cuscuta pentagona* var. *calycina*; parasitic on clover; **a,** flower.

Dodder
Cuscuta spp.
CUSCUTACEAE
Leafless, parasitic perennial
July to September
White
n -

General

Recorded as yielding nectar freely in California and probably does so in Oregon. Not abundant enough to be of significant value. Also known as love vine, love tangle, and coral vine.

Dogbane
Apocynum spp.
APOCYNACEAE
Perennial herb
July and August
Pink, white
n p

Fig. 75 *Apocynum androsaemifolium*

Widely scattered

Several species occur in the Pacific Northwest. Some are heavily worked by bees.

Dog Fennel
Anthemis cotula L.
COMPOSITAE
Annual
May to October
White with yellow center
- ?

Fig. 76 *Anthemis cotula*

General

A common introduced weed species. It is a frequent contaminant of white clover grown for seed. Rarely used by bees as a forage species. Also called May weed.

Fig. 77 *Cornus nuttallii*; a, flower.

Dogwood
Cornus spp.
CORNACEAE
Tree or shrub
Spring
White
- p

Wooded areas

Three dogwood species are known to occur in Oregon. Appear to be of little value to bees as they are rarely seen visiting the blooms.

Fig. 78 *Downingia elegans*

Downingia
Downingia elegans (Lindl.) Torr.
LOBELIACEAE
Annual
July into August
Pale blue
n -

General

Grows in low ground that has been flooded during winter, vernal pools and ditches.

Elderberry, Blue
Sambucus cerulea Raf.
CAPRIFOLIACEAE
Tall shrub
Early spring
Cream
n p

Fig. 79 *Sambucus cerulea*

General in moist places

Important minor plant on the coast. The red elderberry (*S. racemosa* L. var. *arborescens* (T.& G.) Gray) and the black species (*S. racemosa* L. var. *melanocarpa* (Gray) McMinn) appear of little value to bees.

Fig. 80 *Ulmus americana*; a, flowers; b, fruit.

Elm   Tree
Ulmus spp.   Early spring
ULMACEAE   Green   - p

As planted

No elms are native to the Pacific Northwest but many have been planted as shade trees. One of the earliest sources of pollen in abundance.

Fig. 81 *Eriogonum niveum*; a, flower.

Eriogonum, White
Eriogonum niveum Dougl.
POLYGONACEAE
Perennial
June to August
Cream
n p

Sagebrush areas and Ponderosa pine forests east of Cascade crest

In some years, a heavy yielder of light-colored honey. *E. compositum* Dougl. has been observed to yield nectar and pollen near Medford. Also known as snow or wild buckwheat.

Fig. 82 *Oenothera hookeri*

Evening Primrose
Oenothera spp.
ONAGRACEAE
Herb
Late spring into summer
Yellow
? p

Arid regions of eastern Oregon

One species is reported to yield both nectar and pollen in the Hermiston area. Normally only a pollen source, but cultivated varieties produce both nectar and viscid yellow pollen.

Fig. 83 *Anaphalis margaritacea*

Everlasting, Pearly
Anaphalis margaritacea (L.)B.& H.
COMPOSITAE
Perennial herb
Late July and August
White

n p

Abundant in burned-over areas of western Oregon

Conflicting reports as to whether or not it produces surplus honey. Occurs with fireweed, and as fireweed tapers off bees frequently work the everlasting. Nectar sugar measured at 34%.

Fig. 84 *Veratrum californicum*

False Hellebore, White
Veratrum californicum Durand
LILIACEAE
Perennial
Summer
Cream

n p

General in moist to wet areas

Bees have been found dying on this plant in California. Other species of the genus also occur in Oregon. Also known as corn lily.

Farewell-to-Spring
Clarkia amoena (Lehm.) Nels.
& Macbr.
ONAGRACEAE
Annual
June to August
Lavender
- p

Fig. 85 *Clarkia amoena*

Western Oregon

Attractive to honey bees. Also called Summer darling.

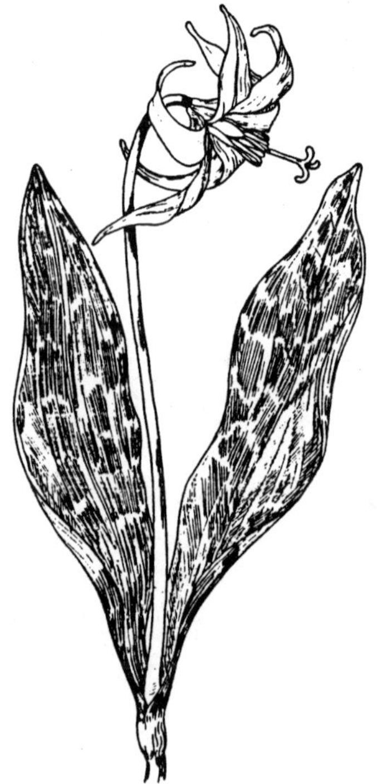

Fawn Lily
Erythronium spp.
LILIACEAE
Perennial herb
Early spring
White, yellow, pink
n p

General in forested areas

E. hendersonii Wats. and *E. oregonum*
App. have been mentioned as nectar and
pollen plants for southwestern Oregon,
but are of minor importance. Also
known as dog tooth violet and Lamb's-
tongue.

Fig. 86 *Erythronium oregonum*

Fig. 87 *Amsinckia intermedia*

Fiddleneck
Amsinckia spp.
BORAGINACEAE
Annual
Spring
Yellow, orange
n p

General

Yields nectar and pollen in California in early spring. Several species in Oregon. Also called amsinckia.

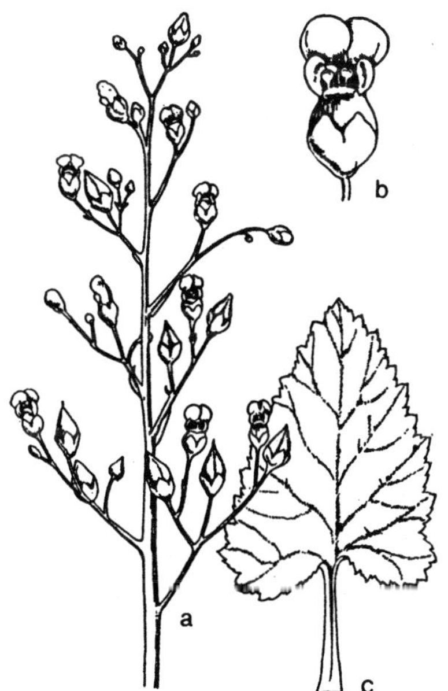

Fig. 88 *Scrophularia californica*;
a, inflorescence; b, flower; c, leaf.

Figwort
Scrophularia spp.
SCROPHULARIACEAE
Perennial herb
May to July
Dark purple
n -

Most common in burned-over sections in western Oregon

Two species *S. californica* C.& S. and *S. lanceolata* Pursh occur. Nectar sugar concentration about half that of fireweed under similar conditions. Pollen collection is not reported.

Fig. 89 *Erodium cicutarium*

Filaree
Erodium cicutarium (L.) L'Her.
GERANIACEAE
Annual
Spring
Rose, purple
n p

General

Good early nectar and pollen source. Also known as alfilaria, heron's bill, stork's bill, etc. Nectar sugars reported as high as 61%. Pollen dark red but fades to brown, easily seen on incoming bees.

Fig. 90 *Corylus avellana*; a, leaves; b, staminate catkins; c, pistillate flower bud.

Filbert, Cultivated hazelnuts
Corylus avellana L.
CORYLACEAE
Shrub to small tree
January and February
Yellow catkins

- p

Commercially grown in the Willamette Valley

One of the earliest pollen sources in the late winter, early spring. Also reported as a source of honeydew.

52

Fig. 91 *Pseudotsuga menziesii*

Fir, Douglas
Pseudotsuga menziesii (Mirb.)
Franco
PINACEAE
Tree
April
Green cones
- p

Western Oregon with a limited occurrence in eastern Oregon

Douglas fir is of interest because of the production of a plant-secreted honeydew known as "fir sugar," reported as common in parts of British Columbia. Douglas fir honey dew uncommon in Oregon.

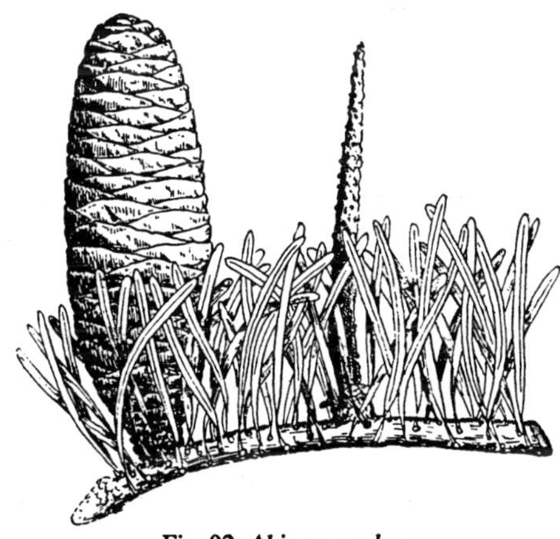

Fig. 92 *Abies concolor*

Fir, White
Abies concolor (G.& G.) Lindl.
PINACEAE
Tree
May and June
Green cones
- ?

Western Oregon

See also Douglas fir. Aphids and scale insects produce honeydew in California from white fir. This is probably true for Oregon.

Fireweed
Epilobium angustifolium L.
ONAGRACEAE
Perennial herb
July and August
Rose, purple
n p

Cascade and Coast Ranges in burned over and clear cut areas

A major source of fine quality honey, but produces irregularly. Fireweed is gradually replaced by a succession of woody plants.

Fig. 93 *Epilobium angustifolium*

Flax
Linum usitatissimum L.
LINACEAE
Annual
July
Pale blue
n p

As planted in western Oregon

Freely visited by honey bees for both nectar and pollen. Of limited value due to infrequent occurrence. The seeds yield linseed oil.

Fig. 94 *Linum usitatissimum*

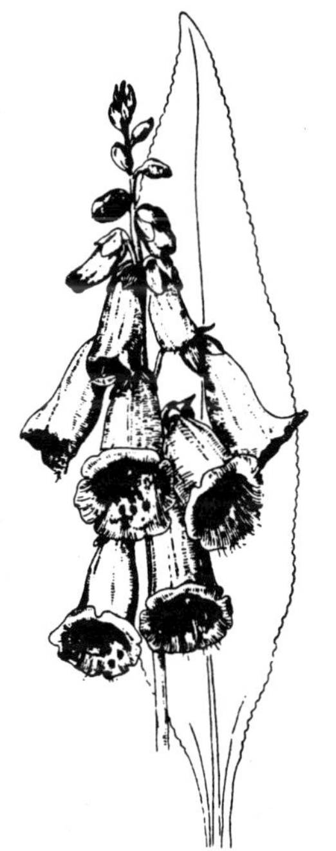

Foxglove
Digitalis purpurea L.
SCROPHULARIACEAE
Biennial herb
June and July
Purple, white
n -

Most common in coastal areas including Coast range mountains. Also cultivated.

An introduced species from Europe. More attractive to bumble bees than honey bees.

Fig. 95 *Digitalis purpurea*

Goldenrod
Solidago spp.
COMPOSITAE
Perennial
July to October
Yellow
n p

General

Of almost no value as a honey plant in the Pacific Northwest. Species occur both west and east of the Cascades but are limited in quantity. The most common western species is *Solidago canadensis* L.

Fig. 96 *Solidago occidentalis*

55

Goldfield
Crocidium multicaule Hook.
COMPOSITAE
Annual
March to May
Yellow
n p

Fig. 97 *Crocidium multicaule*

More common in northeastern Oregon but also in the west.

A good minor plant when found in sufficient density. Also called Gold star. Not worked in the Rogue Valley.

Grape
Vitis spp.
VITACEAE
Perennial vine
Early summer
Greenish yellow
n p

Fig. 98 *Vitis vinifera*

General with a concentration of commercial *V. vinifera* L. in western Oregon

Cultivated varieties are visited for pollen and nectar. Most varieties are self-fertile. Nectar sugars have been measured at 65%.

Fig. 99 *Sorghum halapense*; Johnsongrass

General with a major concentration of commercial seed production in the Willamette Valley

Several species are known to be worked by bees as an occasional pollen source.

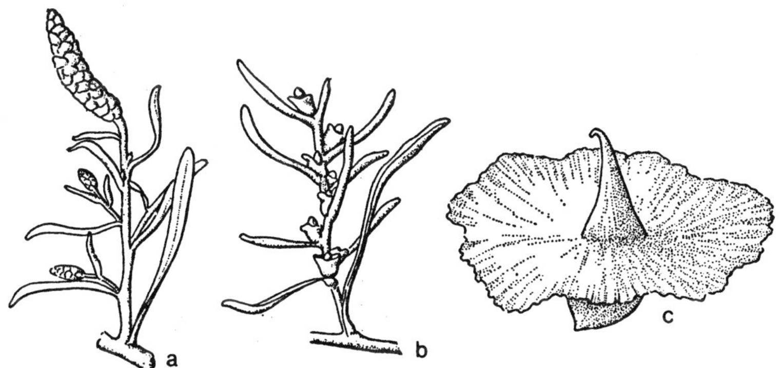

Fig. 100 *Sarcobatus vermiculatus*; a, staminate flower; b, pistillate flower; c, fruit.

Greasewood   Shrub
Sarcobatus vermiculatus (Hook.) Torr.   May to July
CHENOPODIACEAE   Yellow   - p

Alkaline desert areas of eastern Oregon

Pale-colored pollen is freely collected by bees.

Grindelia
Grindelia spp.
COMPOSITAE
Herb
July and August
Yellow
n p

General

Grindelia nana Nutt. is the most common species in both western and eastern Oregon. Bees are not common on it.

Fig. 101 *Grindelia nana*; a, flower.

Fig. 102 *Crataegus douglasii*; a, flowering branchlet; b, fruiting branchlet

Hawthorn
Crataegus spp.
ROSACEAE

Small tree
Late May
White, red   n p

Western Oregon, Wallowa Valley

Several species in Oregon. *C. douglasii* Lindl. is the common western Oregon species. *C. columbiana* How. is found in northeastern Oregon. Good minor source where abundant. Many cultivated species.

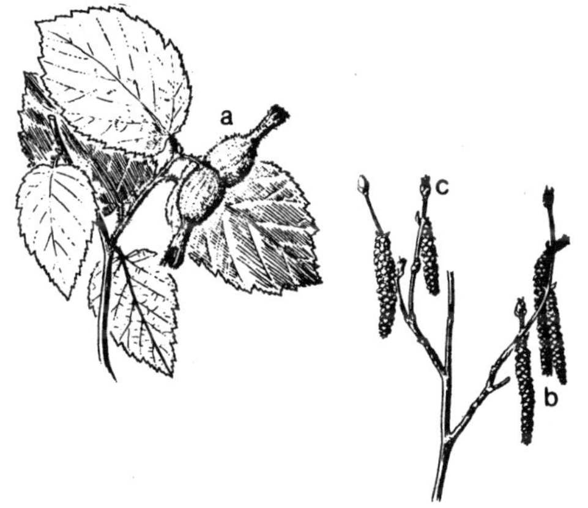

Hazelnut
Corylus cornuta Marsh.
CORYLACEAE
Shrub
January to March
Yellow catkins
- p

Fig. 103 *Corylus cornuta*; a, nuts; b, staminate catkins;
c, pistillate flower buds.

Widespread at lower elevations west of Cascade ridge.

One of the first copious pollen sources available in the late winter.

Heal-all
Prunella vulgaris L.
LABIATAE
Perennial herb
May to September
Purple
n ?

General

Introduced. Also called self-heal and prunella. Bees seldom work it.

Fig. 104 *Prunella vulgaris*

Fig. 105 *Erica carnea*

Heath and Heather
Erica spp. and *Calluna* spp.
ERICACEAE
Low shrub
November to March
White, purple, reds
n p

As planted

Widely planted in gardens. Heavily worked by honey bees when weather permits.

Fig. 106 *Stachys cooleyae*

Hedge-nettle
Stachys spp.
LABIATAE
Perennial or annual
Spring and summer
Red, purple
- -

General

Reported as attractive to bees in other states but not recorded as so in Oregon. Lamb's ears, *S.byzantina* C.Koch, is a cultivated species which supplies nectar.

Holly
Ilex aquifolium L.
AQUIFOLIACEAE
Shrub or tree
Spring
Cream
n p

Fig. 107 *Ilex aquifolium*

As planted

Commercial holly plantings in northern Willamette Valley reported as occasionally yielding surplus honey. Very attractive to bees as a nectar source. Staminate trees supply only pollen.

Hollyhock
Althaea rosea (L.) Cav.
MALVACEAE
Perennial
Summer
White, reds, yellows
n p

As planted

A garden flower introduced from China. Good pollen source when available, and very attractive to bees. Nectar sugar measured at 36%. See Sidalcea.

Fig. 108 *Althaea rosea*

Honeysuckle
Lonicera spp.
CAPRIFOLIACEAE
Shrub or vine
Summer
Yellow, orange, pink
n -

Fig. 109 *Lonicera involucrata* var. *ledebourii*

Mostly in western Oregon

Heavy producers of nectar but honey bees are often unable to reach it. The black twin berry (*L. involucrata* (Rich.) Banks) may be a source of surplus honey in British Columbia.

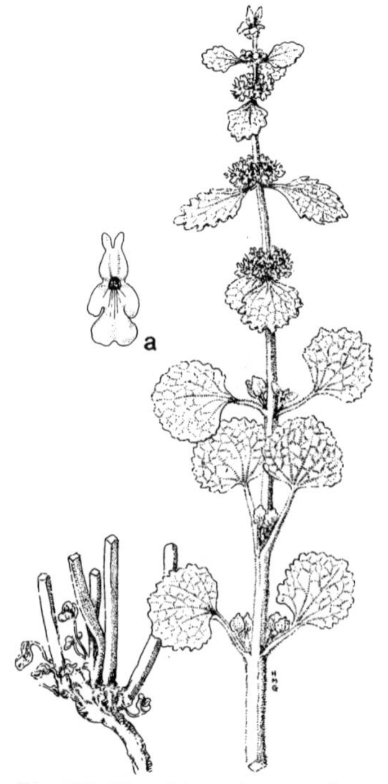

Horehound
Marrubium vulgare L.
LABIATAE
Perennial herb
Summer
White
n -

Widely distributed

Roadsides and barn lots where few other plants will grow. Very attractive to honey bees.

Fig. 110 *Marrubium vulgare*; a, flower.

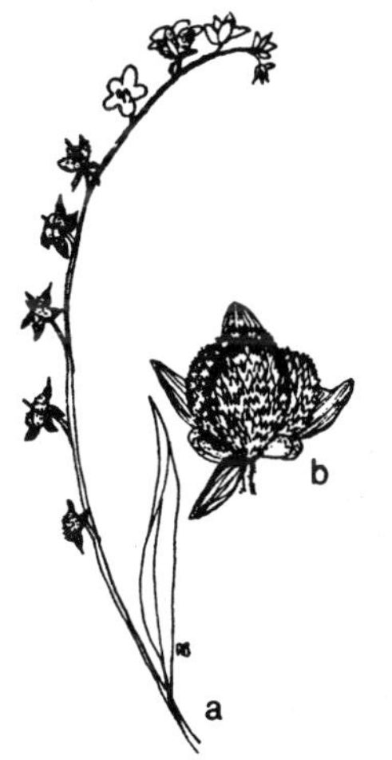

Hound's tongue
Cynoglossum grande Dougl.
BORAGINACEAE
Perennial herb
March to May
Blue
n -

Forested areas at low elevations west of Cascade ridge

Another species of this genus (*C. officinale* L.) which is weedy, is of some importance for honey in Utah. Nectar sugar measured at 48%.

Fig. 111 *Cynoglossum officinale*; a, flowering stem; b, developing nutlets.

Huckleberry
Vaccinium spp.
ERICACEAE
Shrub
Spring
Pink, greenish white
n -

Mountains and coastal

Occasional flows are reported. Several species occur in the Pacific Northwest, frequently in thick patches. Important for spring buildup on the coast.

Fig. 112 *Vaccinium ovatum*; Common evergeen huckleberry

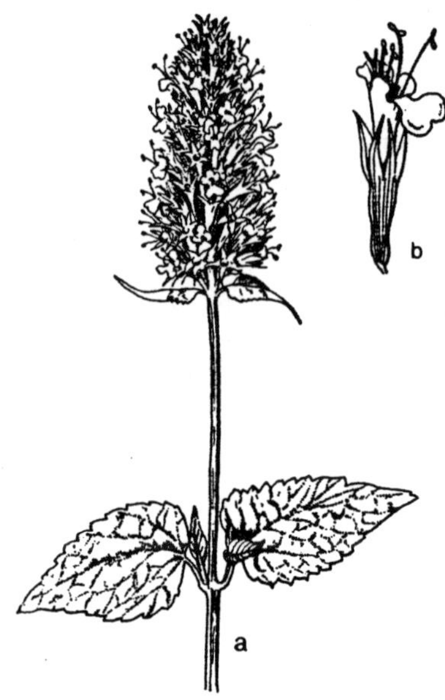

Hyssop, Giant
Agastache urticifolia (Benth.) Kuntze
LABIATAE
Perennial herb
Summer
Purple, white
n -

Hillsides and alpine meadows

Usually out of range of honey bees. Reported as a producer of white honey of fine flavor.

Fig. 113 *Agastache urticifolia*; a, flowering stem; b, flower.

Indian Peach
Oemlaria cerasiformis (T.& G.) Greene
ROSACEAE
Large shrub
March and early April
White
n p

Western Oregon

Also called Indian plum and Oso berry. Bees seldom visit it.

Fig. 114 *Oemlaria cerasiformis*; a, staminate flower; b, pistillate flower.

Fig. 115 *Parthenocissus tricuspidata*

Ivy, Boston
Parthenocissus tricuspidata
(S.& Z.) Planch.
VITACEAE
Vine
Summer
Greenish
n p

As planted

Common as an ornamental and worked freely by bees. Honey reported as dark.
Also known as Japanese creeper.

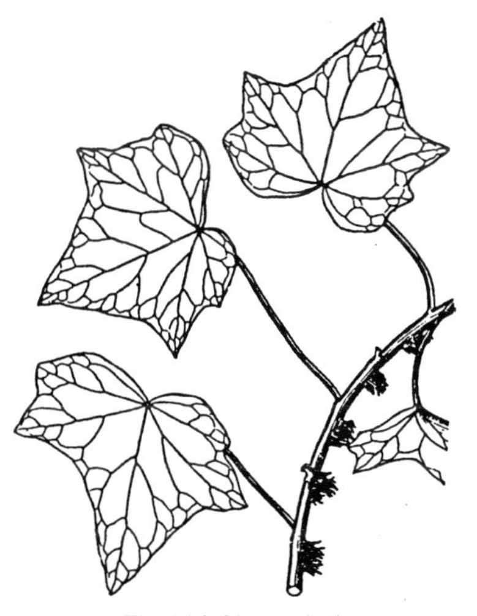

Fig. 116 *Hedera helix*

Ivy, English
Hedera helix L.
ARALIACEAE
Perennial woody vine
September to November
Greenish yellow
n p

As planted

One of the latest nectar and pollen sources. The nectar is very concentrated and
attractive to bees. Flowers on mature growth only.

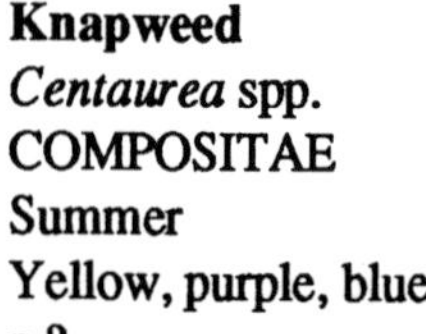

Knapweed
Centaurea spp.
COMPOSITAE
Summer
Yellow, purple, blue
n ?

Central Oregon mainly

Prevalent along dry roadsides and on newly cleared land. Honey has a distinctive flavor.

Fig. 117 *Centaurea jacea*;
Brown knapweed

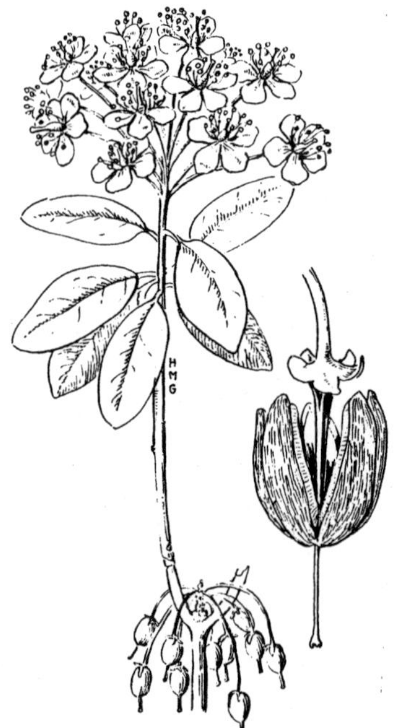

Labrador Tea
Ledum spp.
ERICACEAE
Small shrub
Summer
White
n ?

Bogs of west coast and Blue Mountains

L. glandulosum Nutt. is reported as a source of honey in British Columbia but apparently of little value to bees in Oregon.

Fig. 118 *Ledum glandulosum*

Fig. 119 *Umbellularia californica*; fruiting branchlet.

Laurel, California
Umbellularia californica (H.& A.) Nutt.
LAURACEAE
Tree
Winter to spring
Green
n p

Coast and mountains of Southwest Oregon

Also called bay tree, or locally named Myrtle. Of value as spring stimulant along southwest coast. Honey is dark amber. Copious amounts of pollen.

Fig. 120 *Prunus laurocerasus*

Laurel, English
Prunus laurocerasus L.
ROSACEAE
Shrub or tree
April to May
Cream
n p

As planted

Evergreen shrub widely used in landscaping. Extra-floral nectaries on undersides of leaves also visited by bees.

Laurel, Pale
Kalmia spp.
ERICACEAE
Evergreen shrub
June to September
Pink, purple
? ?

Bogs and mountain meadows

A near relative (*K.latifolia* L.), mountain laurel, which is common in the Atlantic states, is credited with being a source of poisonous honey. Normally out of range of foraging bees.

Fig. 121 *Kalmia polifolia* var. *microphylla*

Laurestinus
Viburnum tinus L.
CAPRIFOLIACEAE
Shrub
Early spring
White
n p

Fig. 122 *Viburnum tinus*

As planted in western Oregon

Introduced and used in western Oregon for landscaping. Not common enough to be of any practical value.

Fig. 123 *Lactuca serriola*; a, fruit.

Lettuce
Lactuca spp.
COMPOSITAE
Annual or perennial herb
Spring and summer
Yellow, blue
n ?

General

Several species of wild lettuce are common weeds. Attracts honey bees to a limited extent but of very minor importance.

Fig. 124 *Ceanothus integerrimus*;
a, flower; b, fruit.

Lilac, Wild
Ceanothus integerrimus H.& A.
RHAMNACEAE
Shrub
April
Blue, white
n p

Coastal woodlands

Good secondary nectar and pollen source where abundant. Cultivated lilacs, *Syringa* species, of the family Oleaceae, are little used by bees.

69

Loco Weed
Astragalus spp.
LEGUMINOSAE
Annual or perennial herb
Spring and summer
Purple, cream, white
n p

Fig. 125 *Astragalus lentiginosus*; a, flowering stem;
b, fruit.

Eastern Oregon

Also known as rattle weed and milk vetch. *A. lentinginosus* Dougl. was found to poison bees in Nevada. Many species are found in Oregon, but there are no reports of injury to bees.

Fig. 126 *Robinia pseudoacacia*; a, flowers; b, fruit.

Locust, Black   Tree
Robinia pseudoacacia L.   May and early June
LEGUMINOSAE   White, pink   n p

As planted. Most common in eastern Oregon

The black locust is far more common and a much better honey plant than the honey locust. Nectar sugar measured as high as 63%. Good source in Eastern Oregon.

Fig. 127 *Gleditsia triacanthos*; a, flowers; b, fruit.

Locust, Honey Tree
Gleditsia triacanthos L. June
LEGUMINOSAE Green n p

As planted. Most common in eastern Oregon

More frequently used as a pollen source. Grown as an ornamental shade tree.
Contrary to the common name, this species is not known as a honey producer.

Loganberries
Rubus ursinus C.& S. var.
loganobaccus (Bailey) Bailey
ROSACEAE
Trailing shrub
Late May into June
White
n p

Fig. 128 *Rubus ursinus* var. *loganobaccus*

Mostly western Oregon

Most members of the genus *Rubus* are freely worked by bees and often produce
surplus honey. Loganberries, youngberries, and boysenberries are hybrids.
Several thousand acres are commercially grown.

Lupine
Lupinus spp.
LEGUMINOSAE
Annual and perennial herbs
Spring
Blue, purple, yellow, white
n p

Fig. 129 *Lupinus bicolor*

General

Many species in Oregon, but rarely reported as honey sources. Bees are known to use *L. bicolor* Dougl. as a nectar source and *L. laxiflorus* Dougl. as pollen source.

Madrone
Arbutus menziesii Pursh
ERICACEAE
Tree
May
White
n p

Fig. 130 *Arbutus menziesii*

Western Oregon especially southwestern areas

Bees may obtain nectar through holes chewed in flowers by other bee species. A thin nectar has been reported (10-20% sugar). Abundant in the Rogue River Valley where it is considered an important major nectar source.

Fig. 131 *Cercocarpus ledifolius*; a, leaves; b, flowers; c, fruiting branch.

Mahogany, Mountain   Shrub
Cercocarpus ledifolius Nutt.   Spring
ROSACEAE   Green   n ?

Mountains of eastern & southern Oregon

Of minor importance.

Mallow
Malva spp.
MALVACEAE
Herb
Early summer
White, pink
n p

Fig. 132 *Malva neglecta*

Widely scattered

Several introduced species. Of some value to bees in other states but of little or
no reported value in Oregon.

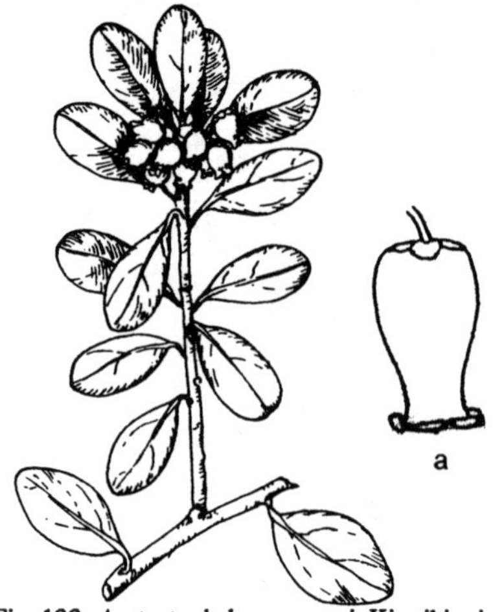

Manzanita
Arctostaphylos spp.
ERICACEAE
Shrub
March and April
White, pink
n p

Fig. 133 *Arctostaphylos uva-ursi*; Kinnikinnic;
a, flower.

Southwestern Oregon and coastal areas

Several species in Oregon, one of which is called kinnikinnic. Some of considerable value in the early spring for buildup. Manzanita species are of major value as nectar sources in the Rogue River Valley and south coast.

Maple, Big-leaf
Acer macrophyllum Pursh
ACERACEAE
Tree
Late March to mid-April
Greenish yellow
n p

Fig. 134 *Acer macrophyllum*

Western Oregon, lower valleys

Provides much pollen and occasionally light amber honey but the early spring weather often handicaps its utilization by bees. Of great value for spring buildup. Nectar sugars range from 35-56%. Also called Oregon Maple.

Fig. 135 *Acer platanoides*

Maple, Norway
Acer platanoides L.
ACERACEAE
Tree
Spring
Greenish yellow
n p

As planted

Occasionally planted as an ornamental shade tree. Nectar sugars reported at 48%. This and other ornamental maples may be important early minor sources where abundant.

Fig. 136 *Acer circinatum*

Maple, Vine
Acer circinatum Pursh
ACERACEAE
Shrub or tree
Late April and May
Red
n p

Western Oregon

Bees visit this plant freely for its abundant nectar and pollen. Occasionally a surplus honey crop is produced. Nectar sugars range from 27 to 58%.

Marigold, Marsh
Caltha spp.
RANUNCULACEAE
Perennial herb
Spring
Yellow, white
n p

Fig. 137 *Caltha biflora*; White marsh marigold

Coastal and mountain bogs

Appears to be of some value to bees but of minor importance.

Meadowfoam
Limnanthes alba L.
LIMNANTHACEAE
Winter annual
May
White
n p

Willamette Valley as a commercial crop

Grown as an oil seed crop on limited acreage. Occasionally yields some honey. Commercial production greatly benefits from honey bee pollination.

Fig. 138 *Limnanthes alba*

Fig. 139 *Asclepias speciosa*

Milkweed
Asclepias spp.
ASCLEPIADACEAE
Perennial herb
Summer
Pink, purple, orange
n -

General but not abundant

Three species in the state. Best known is showy milkweed, *A.speciosa* Torr., with its broad leaves. Insects often trapped by the flowers and die in the blooms. Nectar sugars about 37%.

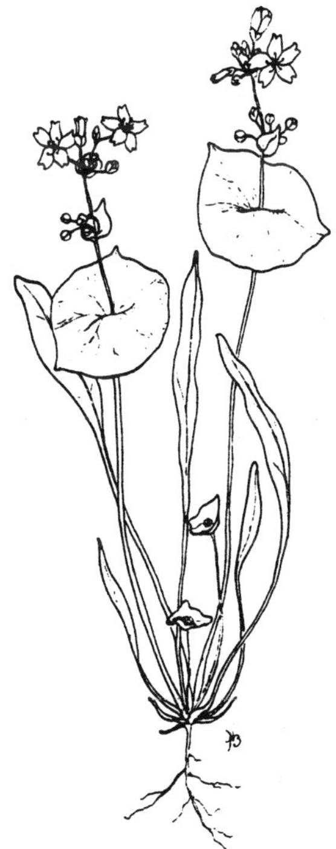

Fig. 140 *Montia perfoliata*

Miner's lettuce
Montia perfoliata (Donn) Howell
PORTULACACEAE
Perennial
March to July
White, pink
n ?

General in moist shady places

Early spring source in some areas.

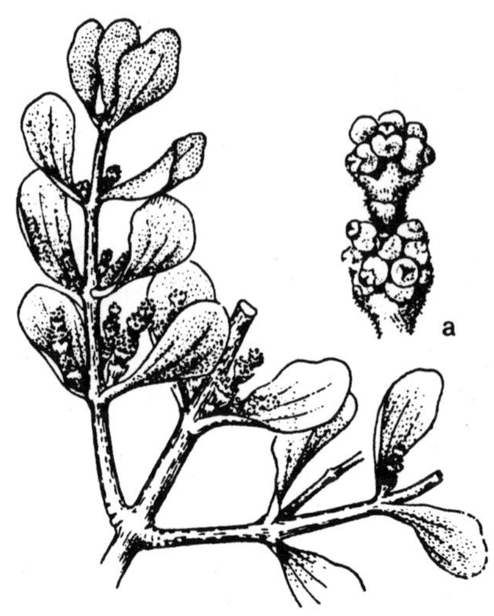

Mistletoe
Phoradendron flavescens (Pursh) Nutt.
LORANTHACEAE
Parasite, mostly on oak
Early spring
Greenish white
- p

Western Oregon, abundant in the
Umpqua and Rogue River Valleys

A very early source of greenish pollen in
limited quantity.

Fig. 141 *Phoradendron flavescens*;
a, inflorescence.

Mock Orange
Philadelphus lewisii Pursh
SAXIFRAGACEAE
Tall bush
May to July
White
? ?

Fig. 142 *Philadelphus lewisii*

General

Of little attraction to bees. Also called syringa. *P. lewisii* var. *gordonianus* is
the western Oregon form.

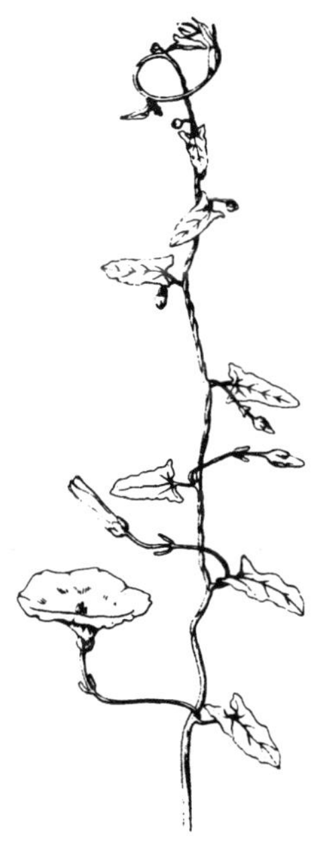

Fig. 143 *Convolvulus arvensis*

Morning Glory
Convolvulus spp.
CONVOLVULACEAE
Perennial trailing herb
Summer
White, blue
n p

General

Of little value in Oregon. *C. arvensis* L. is called Bindweed and gives some nectar and much pollen up to midday.

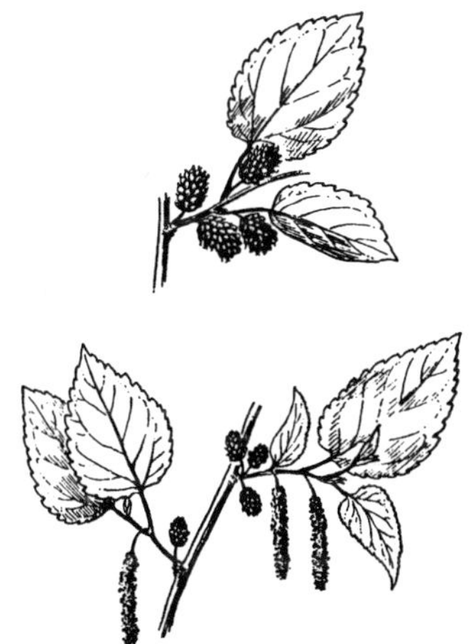

Fig. 144 *Morus alba*

Mulberry
Morus spp.
MORACEAE
Tree
Spring
Green
n ?

As planted

Only occasional trees are found, so it is of little value for honey. Catkins are inconspicuous in bloom.

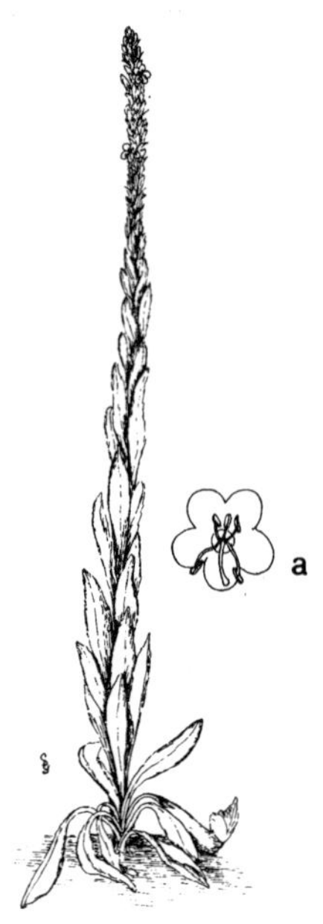

Fig. 145 *Verbascum thapsus*; a, flower.

Mullein
Verbascum thapsus L.
SCROPHULARIACEAE
Biennial
June to August
Yellow
n p

General. Weedy.

Introduced from Eurasia. Produces abundant orange-colored pollen.

Fig. 146 *Verbascum blattaria*; a, flowering
stem; b, stem section.

Mullein, Moth
Verbascum blattaria L.
SCROPHULARIACEAE
Biennial
May to September
Yellow, white
n p

General. Weedy.

Introduced from Eurasia. Of some value for nectar.

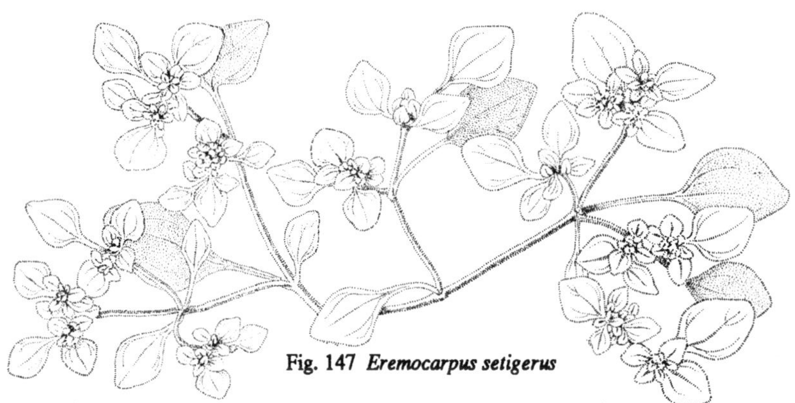

Fig. 147 *Eremocarpus setigerus*

Mullein, Turkey   Prostrate annual
Eremocarpus setigerus (Hook.)Pip.   Summer and fall
EUPHORBIACEAE   Green   n p

General over state in dry ground

Considered more important for pollen than nectar. Also called fish-poison.

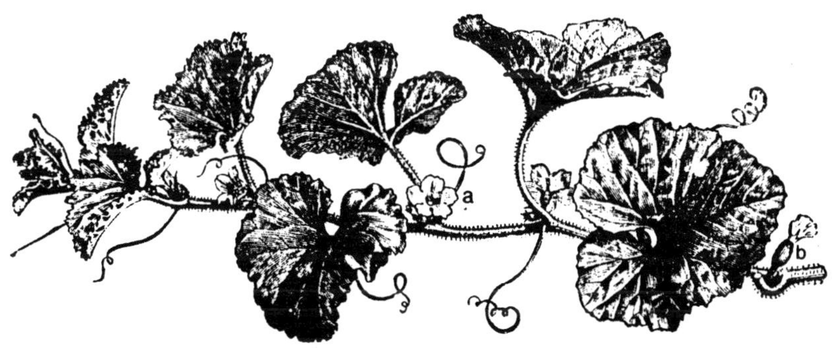

Fig. 148 *Cucumis melo*; a, male flower; b, female flower.

Muskmelon   Annual vine
Cucumis melo L.   Summer
CUCURBITACEAE   Yellow   n p

As planted

Bees are important in pollination, as male and female blossoms are separated on the plant.

Fig. 149 *Brassica nigra*

Mustard, Black
Brassica nigra (L.) Koch
CRUCIFERAE
Annual
May to August
Yellow
n p

Widespread weed but more common in southern Oregon

Blooms later than Common Mustard and is less important.

Fig. 150 *Brassica campestris*

Mustard, Common
Brassica campestris L.
CRUCIFERAE
Annual or biennial
April to June
Yellow
n p

Widespread

Most common of the wild mustards. Produces considerable nectar with 53-72% sugars. Also called Wild Turnip, it frequently attracts bees away from fruit tree bloom.

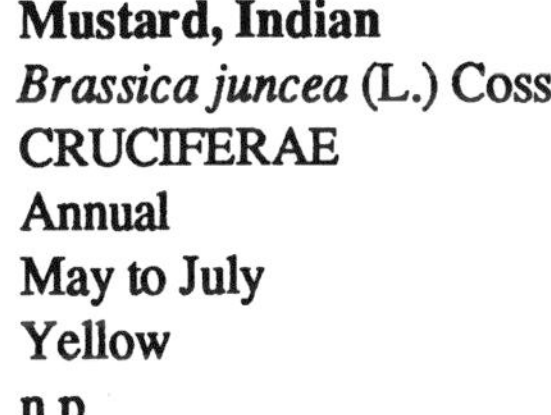

Mustard, Indian
Brassica juncea (L.) Coss.
CRUCIFERAE
Annual
May to July
Yellow
n p

Fig. 151 *Brassica juncea*; a, inflorescence;
b, single flower.

Weedy but not particularly abundant

Also called Charlock.

Mustard, Jim Hill
Sisymbrium altissimum L.
CRUCIFERAE
Annual or biennial
May to September
Cream
n p

Fig. 152 *Sisymbrium altissimum*

Both sides of the Cascades but much more common on the east side

Also called Tumble Mustard. Occasionally yields nectar with about 44%
sugars, and some pollen.

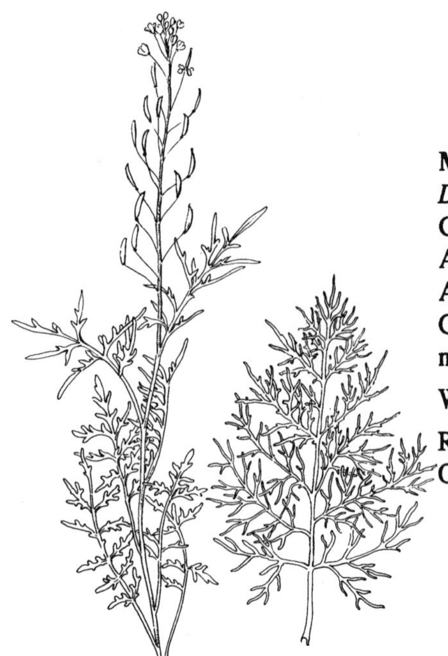

Fig. 153 *Descurainia pinnata*

Mustard, Tansy
Descurainia pinnata (Walt.) Brit.
CRUCIFERAE
Annual
April to July
Greenish yellow
n p

Widespread east of the Cascades

Reported as a source of nectar in central Oregon.

Fig. 154 *Solanum nigrum*; stem with flowers and fruit.

Nightshade
Solanum spp.
SOLANACEAE
Annual or perennial herb
Spring to summer
White, purple
? ?

General

Several species occur in the state. Black nightshade (*Solanum nigrum* L.) is most common. Of little importance.

Ninebark
Physocarpus capitatus (Pursh) Ktze.
ROSACEAE
Tall shrub
May to June
White
n ?

Western Oregon

Bees visit it freely.

Fig. 155 *Physocarpus capitatus*

Fig. 156 a, *Quercus garryana*, White oak; b, *Lithocarpus densiflorus*, Tan oak.

Oak

Quercus spp. and *Lithocarpus* sp.
FAGACEAE

Shrubs and trees
Spring
Yellow catkins   - p

Native in many places and commonly planted

"Oak-dew" may occasionally be collected in late summer from trees heavily parasitized by sawflies. This honeydew sometimes darkens star thistle honey in Rogue Valley.

85

Ocean Spray
Holodiscus discolor (Pursh) Maxim.
ROSACEAE
Tall shrub
June to July
Cream
- -

Western and northeastern Oregon

Rarely visited by honeybees.

Fig. 157 *Holodiscus discolor*

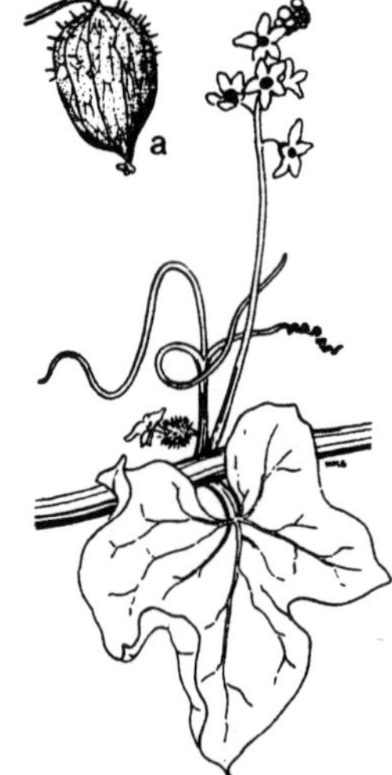

Old-man-in-the-ground
Marah oreganus (T.& G.) How.
CUCURBITACEAE
Climbing perennial herb
April and June
White
n p

Fig. 158 *Marah oreganus*; a, fruit.

Mostly western Oregon

Also called wild cucumber. Of little value to honey bees.

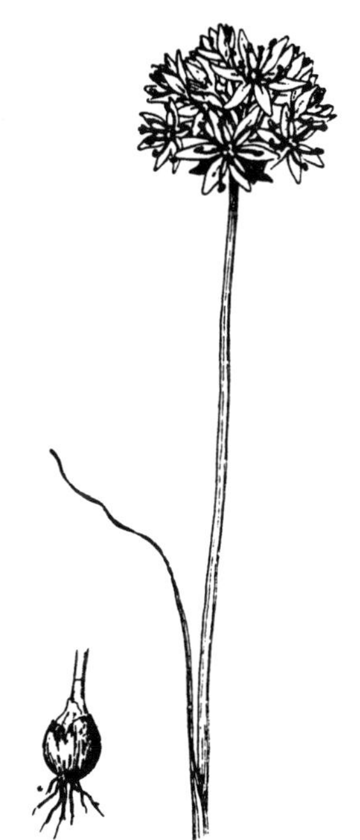

Onion
Allium spp.
LILIACEAE
Mostly perennial
Spring
White, pink, reds
n p

Native species occur throughout the state.

The cultivated onion (*A. cepa* L.) is a good honey plant when grown for seed. Honey bees are effective pollinators of onions. Wild species are occasionally of value in eastern Oregon.

Fig. 159 *Allium amplectens*; Wild onion

Oregon Grape
Berberis spp.
BERBERIDACEAE
Shrub
April and May
Yellow
n p

Fig. 160 *Berberis aquifolium*

Especially on hills

Of three species common in the state, best known is *B. aquifolium* Pursh, Oregon's state flower. This species is of secondary importance. Nectar sugar concentration varies 40-54%. May be classified as *Mahonia* by some taxonomists.

Parsley
Petroselinum crispum (Mill.)
Nym.
UMBELLIFERAE
Biennial herb
Late June through July
Greenish yellow
n p

Fig. 161 *Petroselinum crispum*; a, umbel; b, fruit.

As planted, especially in Willamette Valley

Commercially grown for seed on limited acreage in western Oregon. Yields greatly increased by honey bee pollination. Very attractive to honey bees.

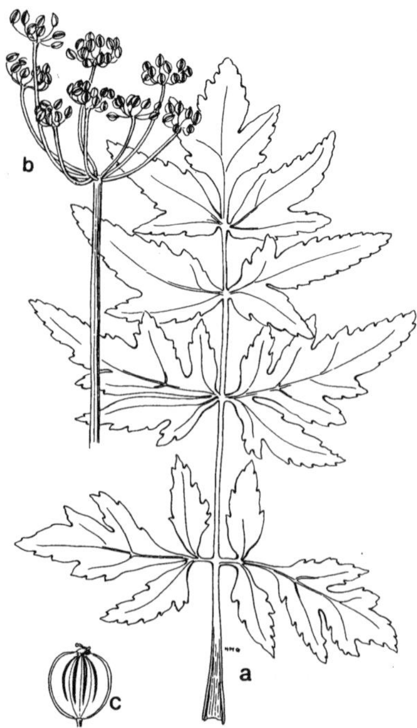

Parsnip
Pastinaca sativa L.
UMBELLIFERAE
Biennial
Spring
Yellow
n ?

As planted for seed

A good honey producer when grown for seed. Occasionally escaped and established as a roadside weed.

Fig. 162 *Pastinaca sativa*; a, lower leaf; b, umbel of fruits; c, fruit.

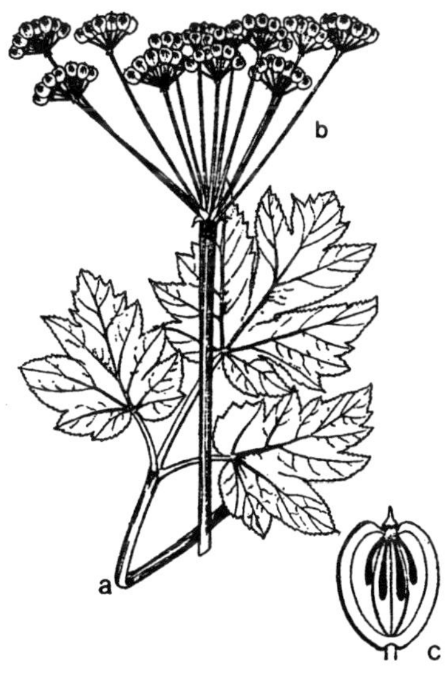

Parsnip, Cow
Heracleum lanatum Michx.
UMBELLIFERAE
Perennial herb
Late May to August
White
n p

General in moist areas

Of minor value.

Fig. 163 *Heracleum lanatum*; a, leaf;
b, umbel; c, fruit.

Pea
Lathyrus spp.
LEGUMINOSAE
Annual or perennial vine
Spring and early summer
Yellow, purple, red
n p

As planted

Peas are of little value for nectar, but
supply some pollen. Native species occur
throughout the state. Nectar sugars in
Austrian Peas about 50%.

Fig. 164 *Lathyrus latifolius*; Wild sweet
pea

Peach
Prunus persica (L.) Batsch.
ROSACEAE
Tree
Early spring
Pink
n p

Fig. 165 *Prunus persica*

Willamette and Rogue River Valleys, and parts of eastern Oregon

Nearly all peaches are self-fertile, but a better set is obtained when bees are present. Nectar sugars average 30%.

Fig. 166 *Pyrus communis*

Pear   Tree
Pyrus communis L.   April and May
ROSACEAE   White   n p

Hood River, Willamette, Rogue River Valleys

Very valuable commercially. Most leading varieties are self-sterile and set larger crops of better fruit when suitable pollinizers are present and bees are available. Low nectar sugars, 5-25%.

Fig. 167 *Mentha pulegium*; a, inflorescence; b, stem and root.

Pennyroyal
Mentha pulegium L.
LABIATAE
Perennial herb
Summer to fall
Rose, lavender
n ?

Western Oregon

A naturalized weed. Reported as a valuable honey plant in California.

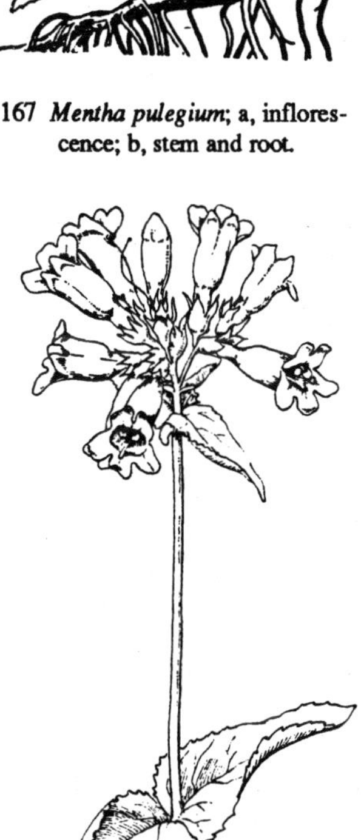

Fig. 168 *Penstemon serrulatus*

Penstemon
Penstemon spp.
SCROPHULARIACEAE
Mostly perennial herbs
Late spring to summer
Cream, blue, purple
n ?

Alpine meadows and rock gardens

Very attractive to bees but the plants are too few where bees are available. Nectar sugars about 35%.

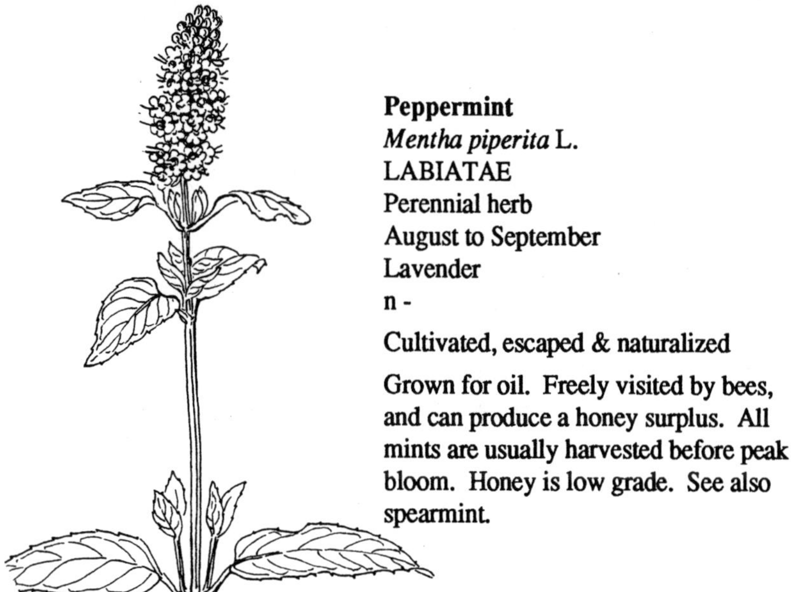

Fig. 169 *Mentha piperita*

Peppermint
Mentha piperita L.
LABIATAE
Perennial herb
August to September
Lavender
n -

Cultivated, escaped & naturalized

Grown for oil. Freely visited by bees, and can produce a honey surplus. All mints are usually harvested before peak bloom. Honey is low grade. See also spearmint.

Fig. 170 *Phacelia californica*; a, leaf;
b, flowering stem.

Phacelia
Phacelia spp.
HYDROPHYLLACEAE
Herbaceous perennial or annual
Late spring, early summer
Blue, purple, cream, yellow
n p

General where moist

One or more species of the genus can be found throughout the state. Important minor source of nectar and pollen. Commonly called fiddleneck.

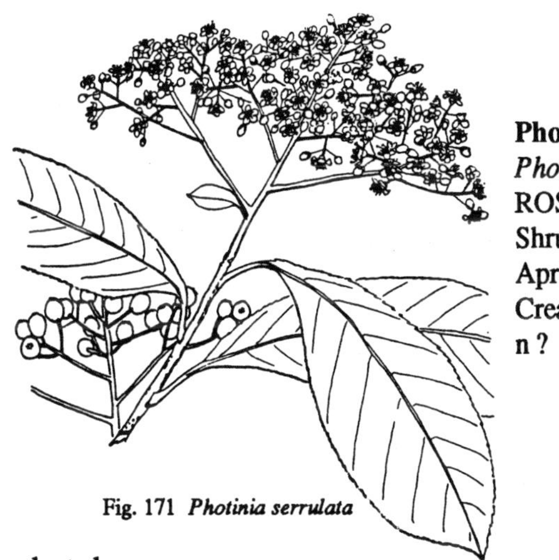

Photinia
Photinia spp.
ROSACEAE
Shrub
April and May
Cream
n ?

Fig. 171 *Photinia serrulata*

As planted

Laurel-like shrub with bright red new growth. Used extensively in home landscaping and along highways.

Pine
Pinus spp.
PINACEAE
Tree
Various
Yellow male cones
- p

Fig. 172 *Pinus ponderosa*

General

Bees collect pollen in quantity from the catkins. Occasionally pines are a source of honeydew.

Fig. 173 *Plagiobothrys tenellus*; a, nutlet.

Plagiobothrys
Plagiobothrys tenellus (Nutt.)
Gray
BORAGINACEAE
Annual
April to June
White
n ?

Widespread

Also called small popcorn flower. Grows in ground that is very wet in winter and baked dry in summer. Nectar sugars about 57%.

Fig. 174 *Plantago lanceolata*

Plantain, English
Plantago lanceolata L.
PLANTAGINACEAE
Perennial herb
April to August
Cream
- p

General in lawns, etc.

An introduced weed commonly called buckhorn. The several native, broadleaved species are less visited by bees.

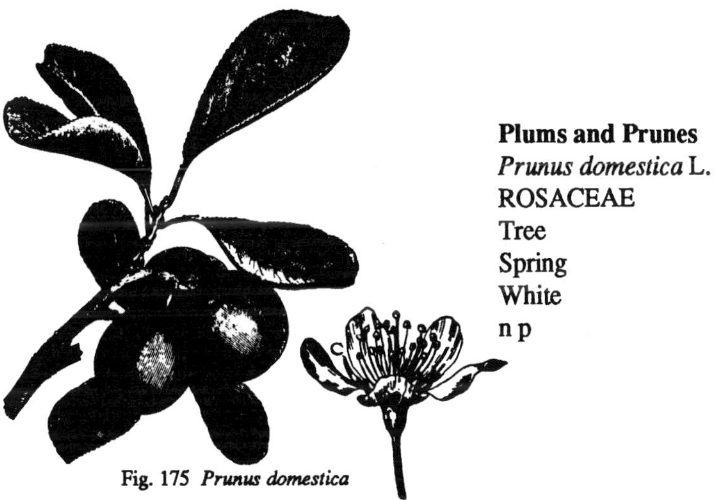

Fig. 175 *Prunus domestica*

Plums and Prunes
Prunus domestica L.
ROSACEAE
Tree
Spring
White
n p

Willamette Valley, Douglas County, Umatilla County

Prunes are a form of plums. Bees are beneficial to ensure efficient pollination and a larger crop. Plums and prunes secrete nectar (with 14-44% sugars) at low temperatures.

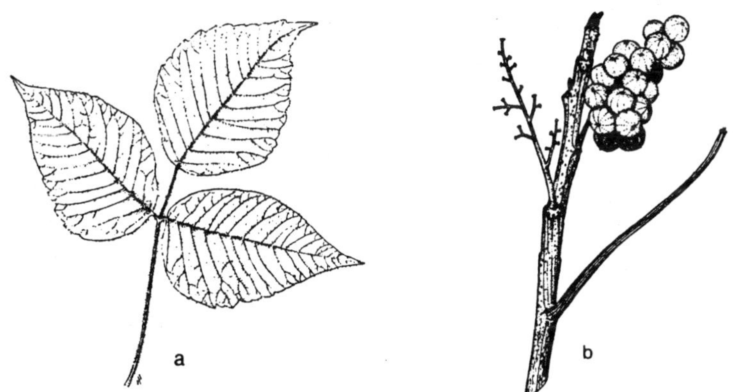

Fig. 176 *Rhus radicans*; a, leaf; b, fruiting stem.

Poison Ivy
Rhus radicans L.
ANACARDIACEAE

Shrub or climber
April to July
Cream   n p

Parts of eastern Oregon

It has not been reported of much value to bees in Oregon.

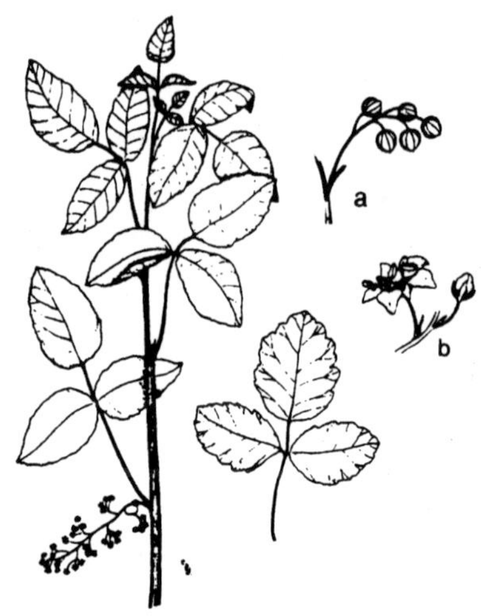

Poison Oak
Rhus diversiloba T.& G.
ANACARDIACEAE
Shrub or climber
May to July
Greenish white

n p

Western Oregon

Producer of light honey in areas where it grows in sufficient abundance. The honey is reputed to be of some therapeutic value to those sensitive to poisoning by the plant.

Fig. 177 *Rhus diversiloba*; a, fruit; b, flower and bud.

Polygonum
Polygonum spp.
POLYGONACEAE
Annual or perennial herb
Summer and fall
Red, white, green
n -

Fig. 178 *Polygonum hydropiperoides*; a, inflorescence; b, flowers; c, leaves.

Found throughout the state

Both *P. hydropiperoides* Michx. and *P. spergulariaeforme* Meisn. have been reported as nectar plants. Honey from the latter is dark amber and of fine quality. Of little value to bees in general.

Poppy, California
Eschscholzia californica Cham.
PAPAVERACEAE
Perennial herb
Spring and summer
Yellow, orange
- p

Fig. 179 *Eschscholzia californica*

Escaped in Western Oregon, also cultivated

An important source of golden pollen. Other cultivated poppies, *Papaver* species, produce high quality pollen which is very attractive to bees.

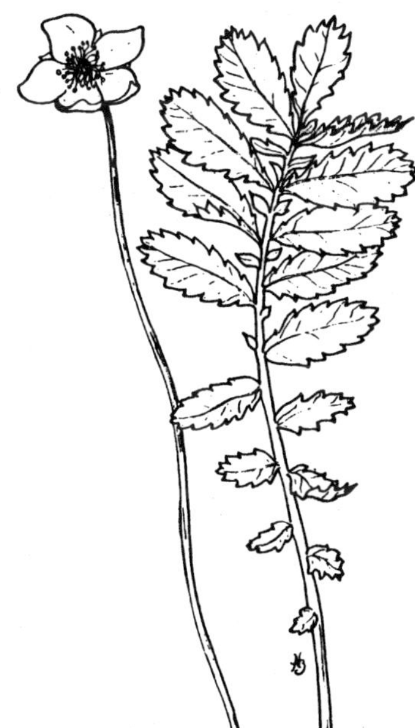

Potentilla
Potentilla spp.
ROSACEAE
Perennial herb or shrub
Early summer
Yellow, white, purple
n p

General

Most species are not very attractive to bees.

Fig. 180 *Potentilla pacifica*; Silverweed

Privet
Ligustrum spp.
OLEACEAE
Tall shrub
Spring
Cream
n p

Fig. 181 *Ligustrum vulgare*

As planted

Planted extensively as a shrub for hedges. Not common enough to be of much value.

Psoralea
Psoralea lanceolata Pursh
LEGUMINOSAE
Perennial herb
April to September
White, pale blue
n ?

Fig. 182 *Psoralea lanceolata*

Eastern Oregon

Also called lance-leaf scurf pea. California tea, *P. physodes* Dougl., is reported as a nectar source in southwestern Oregon.

Pumpkin, Squash, Gourd
Cucurbita pepo L.
CUCURBITACEAE
Annual vine
June to September
Yellow
n p

Fig. 183 *Cucurbita pepo*; a, male flower; b, female
flower.

As planted

These plants supply a liberal amount of both pollen and nectar. The honey is
amber and of poor grade. Commercial crops of pumpkin and squash greatly
benefit from honey bee pollination.

Pussy Paws
Spraguea umbellata Torr.
PORTULACACEAE
Perennial herb
Summer
Rose
n ?

Fig. 184 *Spraguea umbellata*

Dry soil in Cascade Mountains and eastward, also in Southwest Oregon
Probably of little value.

Pyracantha
Pyracantha spp.
ROSACEAE
Shrub
May and June
Cream
n p

Fig. 185 *Pyracantha koidzumii*

As planted

Widely planted fence or hedge plant. Heavily worked by bees. Also called Firethorn.

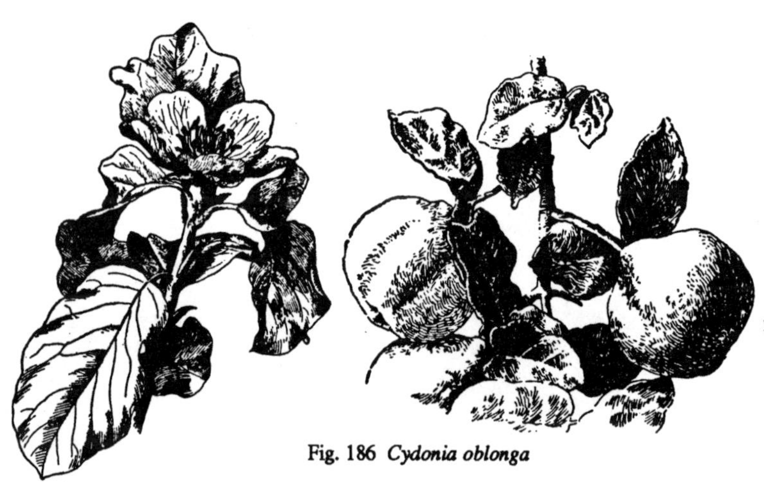

Fig. 186 *Cydonia oblonga*

Quince   Tree
Cydonia oblonga Mill.   Spring
ROSACEAE   White, pink   n p

Very limited in Oregon

Flowers of this fruit tree are freely visited by bees.

Fig. 187 *Chrysothamnus nauseosus*;
a, floret.

Rabbit Brush
Chrysothamnus spp.
COMPOSITAE
Shrub
August to early September
Yellow
n p

Chiefly in Eastern Oregon

Most common species is *C. nauseosus* (Pall.) Britton, which apparently requires very little moisture and can tolerate alkaline conditions. Honey is amber and crystallizes rapidly.

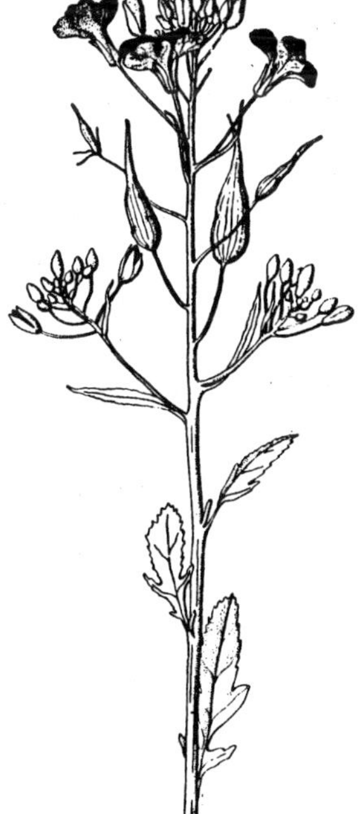

Fig. 188 *Raphanus sativus*

Radish
Raphanus sativus L.
CRUCIFERAE
Annual or biennial
May to July
Yellow
n p

Weedy and as planted

Very attractive to bees. Commercially grown on limited acreage in the Willamette Valley and Central Oregon. Honey is extra light amber, good flavor, and granulates rapidly.

Rape
Brassica napus L.
CRUCIFERAE
Annual or biennial
May
Yellow
n p

Fig. 189 *Brassica napus*

Primarily as planted in the Willamette Valley

This "domesticated" mustard is planted commercially for seed in Western Oregon and is very attractive to bees. Abundant nectar and pollen produced. Nectar sugar concentration is 50-73%.

Fig. 190 *Rubus* spp.

Raspberry	Shrub
Rubus spp.	May and June
ROSACEAE	White n p

Commercially grown in northern Willamette Valley

An occasional source of fine surplus honey, and very attractive to bees. Honey-bee pollination is considered important in the formation of well-shaped fruit and a good crop.

Red Maids
Calandrinia ciliata (R. & P.) DC.
PORTULACACEAE
Annual
April to early summer
Red
n p

Weedy in cultivated areas

Considerable yellow-amber nectar of 48% sugar concentration is produced. Pollen is orange in color.

Fig. 191 *Calandrinia ciliata*

Rhododendron
Rhododendron spp.
ERICACEAE
Tall shrub
May and June
Pinks, white, purple, yellow
n ?

Fig. 192 *Rhododendron macrophyllum*

Moist, cool mountain habitat to sea level

Several species are found in coastal and montane areas, and many species are cultivated in gardens. Rarely visited by honey bees but worked by bumble bees.

Rocky Mountain Bee Plant
Cleome serrulata Pursh
CAPPARIDACEAE
Annual
June to August
Purple
n p

Eastern Oregon

Grows best in undisturbed, arid, alkaline soil. Nectar sugar concentration as high as 22%. May provide a surplus of light, strong flavored honey. Bees also gather greenish pollen from flowers.

Fig. 193 *Cleome serrulata*; a, flower.

Rose, Wild
Rosa spp.
ROSACEAE
Shrub
May and early June
Pink
n p

Fig. 194 *Rosa eglanteria*; Sweetbriar rose; a, fruit.

General

Sweet briar rose is especially abundant in the Roseburg area and in the Willamette Valley as an escape. Very little nectar is collected.

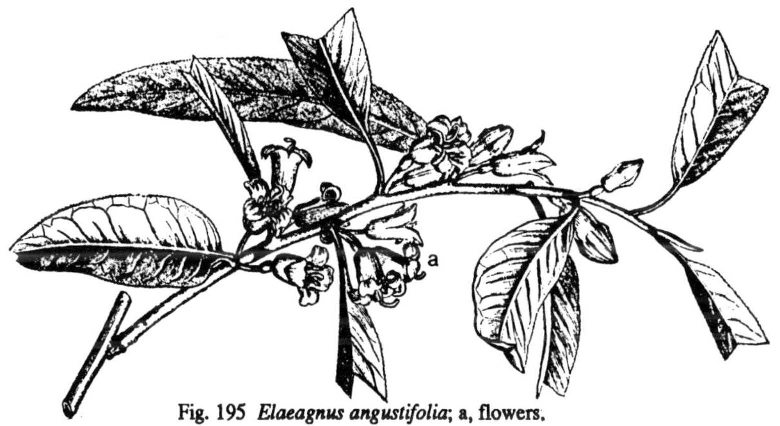

Fig. 195 *Elaeagnus angustifolia*; a, flowers.

Russian Olive	Tree
Elaeagnus angustifolia L.	Late June
ELAEAGNACEAE	Yellow n p

Eastern Oregon as planted

Also called silverberry. Planted as a shade tree along highways in Eastern Oregon. Nectar sugar concentration about 41%.

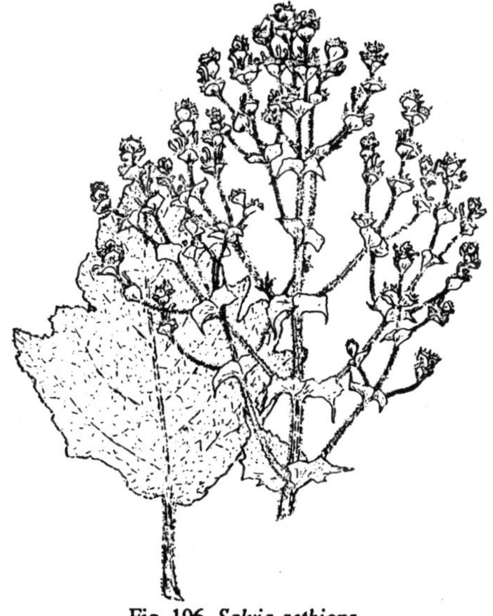

Sage
Salvia spp.
LABIATAE
Biennial, perennial or shrubby
Early summer
Blue, purple, white
n -

Fig. 196 *Salvia aethiops*

Mostly Eastern Oregon

Desert sage, *S. dorrii* (Kell.)Abrams, native to Oregon in sandy and gravelly soils, yields only in moist years. Mediterranean sage, *S. aethiops* L., is established in eastern Oregon.

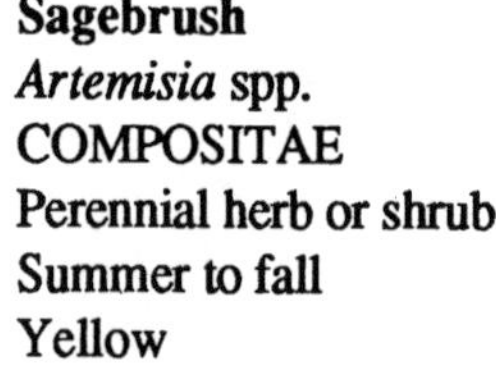

Sagebrush
Artemisia spp.
COMPOSITAE
Perennial herb or shrub
Summer to fall
Yellow
- p

Eastern Oregon

Sagebrush produces a quantity of pollen. It should be clearly distinguished from the true sages of the genus *Salvia*.

Fig. 197 *Artemisia tridentata*; Big sagebrush; a, flowering stem; b, leafy twig.

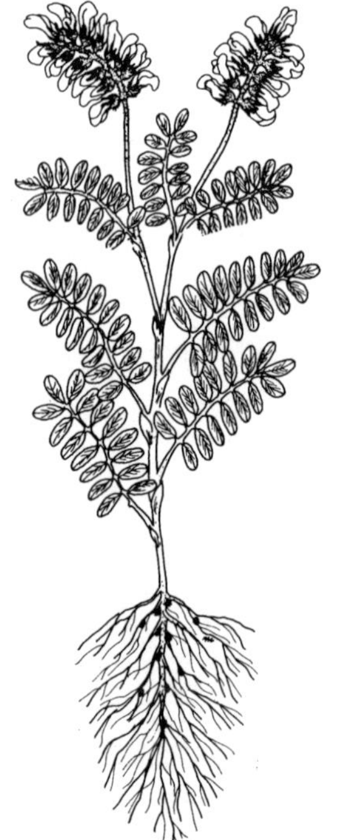

Sainfoin
Onobrychis viciifolia Scop.
LEGUMINOSAE
Perennial herb
June to August
Pink
n ?

As planted

Limited in Oregon. Also called Holy clover. Nectar contains about 55% sugars. Grown in Europe as a forage crop on poor land.

Fig. 198 *Onobrychis viciifolia*

Salal
Gaultheria shallon Pursh
ERICACEAE
Low shrub
May to July
White, pink
n -

Coast to west slopes of the Cascades. Also cultivated.

Flowers produce an abundant supply of nectar. The honey is light amber in color and good flavored. Yields some honey on the west side of the Cascades where salal is thick and abundant.

Fig. 199 *Gaultheria shallon*

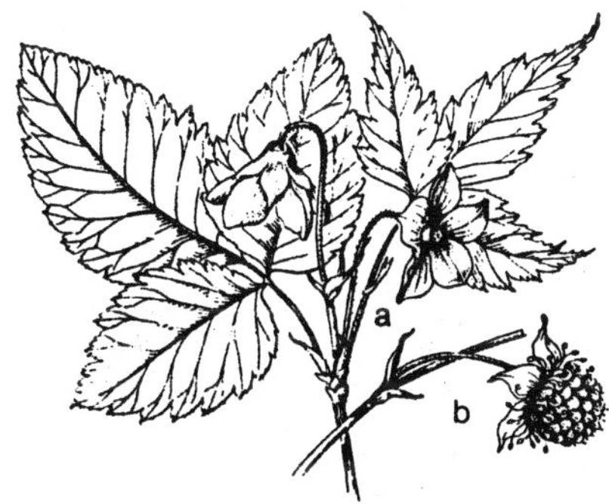

Salmonberry
Rubus spectabilis Pursh
ROSACEAE
Shrub
March to June
Red
n p

Fig. 200 *Rubus spectabilis*; a, flowering branch; b, fruit.

Western Oregon, uncommon in east

In swamps and along streams. An important minor plant but too early for a surplus.

Scots Broom
Cytisus scoparius Link
LEGUMINOSAE
Shrub
May and June
Yellow
? p

Western Oregon

An introduced ornamental shrub widely escaped and a pest in many sections. Some years well worked for abundant yellow pollen. Domestic variegated species more attractive to bees.

Fig. 201 *Cytisus scoparius*

Fig. 202 *Amelanchier alnifolia*; a, leaves; b, flowers.

Service Berry   Tall shrub or small tree
Amelanchier spp.   April to July
ROSACEAE   White   ? p

General

Three species in the state, apparently useful to bees only for pollen. Also called servies berry.

Sidalcea

Sidalcea spp.
MALVACEAE
Annual or perennial herb
Late spring and summer
Pink, purple, white
n p

General

Also called wild hollyhock. Common along roadsides. Very little nectar collected.

Fig. 203 *Sidalcea campestris*; Tall wild
hollyhock

Silene

Silene spp.
CARYOPHYLLACEAE
Annual or perennial herb
Spring to summer
White, pink
? p

General

An unidentified species has been reported as a source of pollen.

Fig. 204 *Silene hookeri*; Indian pink

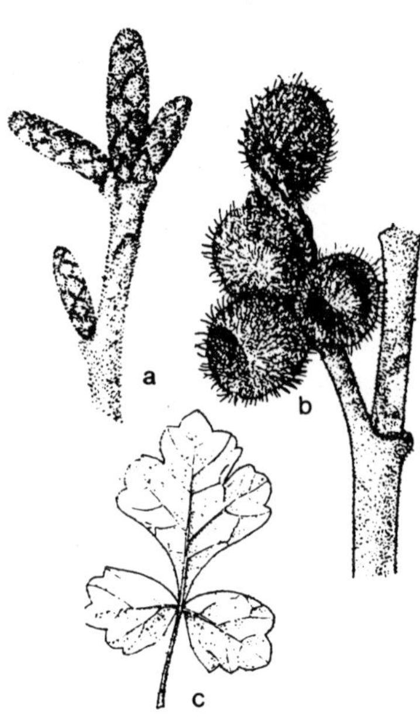

Skunk Bush
Rhus trilobata Nutt.
ANACARDIACEAE
Shrub
May to July
Yellow
n p

Southeast and southwest Oregon

Very attractive to bees. Also called squaw bush.

Fig. 205 *Rhus trilobata*; a, flower buds; b, fruit; c, leaf.

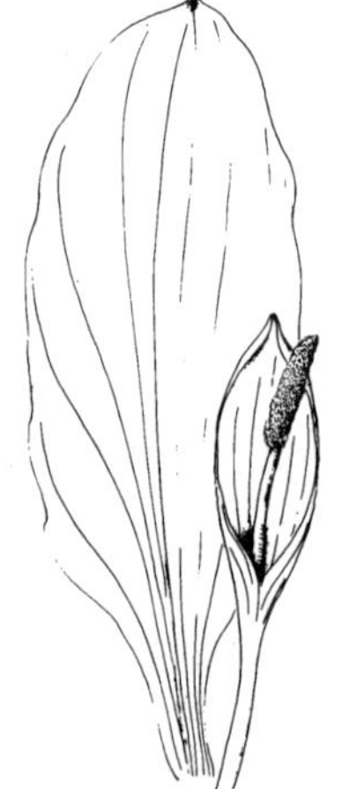

Skunk Cabbage
Lysichitum americanum Hult. & St.J.
ARACEAE
Perennial herb
April to July
Yellow
- p

Fig. 206 *Lysichitum americanum*

Chiefly Western Oregon swamps

Much pollen produced, imparting a pungency to honey. This is not the same plant as the skunk cabbage of the east (*Symplocarpus foetidus*).

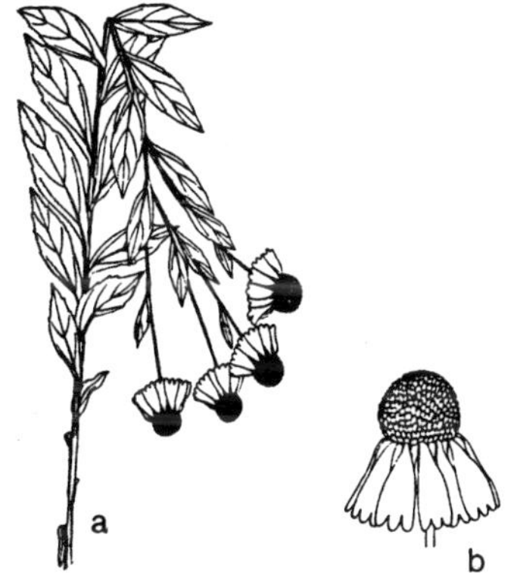

Sneezeweed
Helenium autumnale L.
COMPOSITAE
Perennial
July to September
Yellow
n p

Fig. 207 *Helenium autumnale*; a, flowering stem;
b, flower.

Chiefly along Snake, Columbia and Klamath Rivers

Not reported as being visited by bees in Oregon, but a good honey plant in north central states. The honey is somewhat bitter.

Snowberry
Symphoricarpos spp.
CAPRIFOLIACEAE
Shrub
May to August
White, pink
n -

Western Oregon and Blue Mountains

Common species is *S. albus* (L.) Blake, abundant in open woods of western Oregon, which may yield a light flow June to August. Nectar sugar concentration 47-58%. Pollen collection not observed.

Fig. 208 *Symphoricarpos albus*; a, fruit.

Soap Root
Chlorogalum pomeridianum
(DC.) Kunth
LILIACEAE
Perennial herb
Summer
White with purple veins
n p

Fig. 209 *Chlorogalum pomeridianum;* a, basal leaves
and bulb; b, flowering stem; c, inflorescence.

Roseburg and south

Probably of considerable value where common. The blossoms open in late
afternoon, remaining only for one night. Also called soap plant.

Sorrel, Sheep
Rumex acetosella L.
POLYGONACEAE
Perennial herb
May to August
Red pistils, yellow anthers
- p

General

A common weed in acid soil. Separate
male and female plants. Also called red
sorrel.

Fig. 210 *Rumex acetosella*

Sorrel, Wood
Oxalis spp.
OXALIDACEAE
Low herb
April to October
White, lavender, yellow
n p

General

Common in shady woods. Cultivated form is also visited for pollen.

Fig. 211 *Oxalis oregona*; Wood sorrel

Spearmint
Mentha spicata L.
LABIATAE
Perennial herb
June to August
Blue, white
n ?

Fig. 212 *Mentha spicata*

Damp ground where escaped from cultivation.

Bees work it freely, primarily for nectar. Grown in Oregon for oil production, but on a smaller acreage than is peppermint. See also peppermint.

Spikeweed
Hemizonia pungens (H.& A.)
T.& G.
COMPOSITAE
Annual
July to September
Yellow
n p

Fig. 213 *Hemizonia pungens*; a, basal leaf.

Common in Umatilla Co. Occasional occurrence elsewhere.

Little is known of the value of this uncommon plant in Oregon. Source of nectar and pollen in California central valley. Also called alkali weed.

Fig. 214 a, *Spiraea douglasii*; b, *S. betulifolia*.

Spiraea Shrub
Spiraea spp. Spring and early summer
ROSACEAE Pink, white ? p

Native species in Willamette Valley, Cascade and Blue Mountains.

A minor source of grey pollen. There are also many cultivated species of *Spiraea*.

Spring Gold
Lomatium spp.
UMBELLIFERAE
Perennial herb
Spring and summer
Yellow
? p
General

Some species are a source of pollen and possibly also nectar. Also called parsley.

Fig. 215 *Lomatium utriculatum*

Spruce
Picea spp.
PINACEAE
Tree
May to July
Green cones
- -

Fig. 216 *Picea sitchensis*

In mountains and along coast

Has been known to produce large amounts of honeydew, but none of the Oregon species, native or introduced, have been reported as honeydew sources.

St. John's-Wort
Hypericum perforatum L.
HYPERICACEAE
Perennial
June to July
Yellow
n p

Abundant in western Oregon, also found east of the Cascades

An introduced weed of minor importance for nectar, but good source of pollen. Also called Klamath weed, tipton weed and goat weed.

Fig. 217 *Hypericum perforatum*

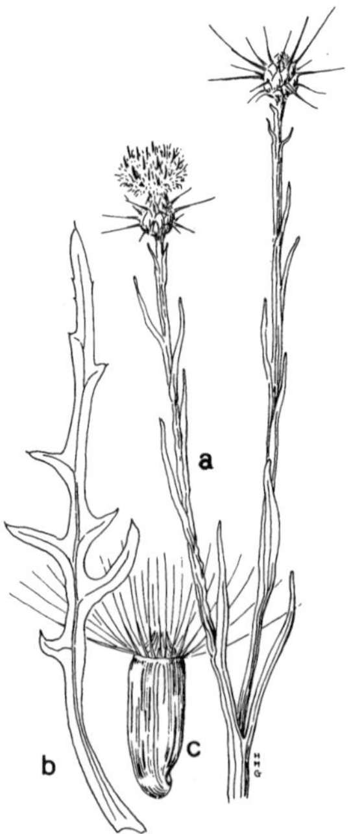

Star Thistle, Yellow
Centaurea solstitialis L.
COMPOSITAE
Annual or biennial
July to September
Yellow
n p

From Roseburg south and in eastern Oregon

An introduced weed from Europe which has become a locally important source of nectar. Honey is yellow with excellent flavor. Nectar sugar concentration 37%. See also Bachelor Button.

Fig. 218 *Centaurea solstitialis*; a, upper branches; b, lower leaf; c, fruit.

Fig. 219 *Spergula arvensis*; a, flowering
stem; b, habit.

Starwort
Spergula arvensis L.
CARYOPHYLLACEAE
Annual
March to October
White
n ?

Chiefly west of the Cascades

Also called corn spurry. A common
garden weed from Europe. Reported as
yielding some nectar in Clatsop County.

Fig. 220 *Fragaria* spp.

Strawberry
Fragaria spp.
ROSACEAE
Perennial herb
Spring
White
n p

Moist places and as planted

Of minor importance for bees. Many commercial cultivars of strawberry are
greatly benefited by the pollinating activities of bees, but the flowers are not
very attractive to·bees.

Sumac, Smooth
Rhus glabra L.
ANACARDIACEAE
Shrub
April to July
Greenish white
n p

Moist ground in canyons of eastern Oregon

Considered important in eastern states but scarce and of little value in Oregon. Poison oak and skunk bush belong to the same genus.

Fig. 221 *Rhus glabra*

Sunflower
Helianthus annuus L.
COMPOSITAE
Tall annual
Summer
Yellow
n p

Fig. 222 *Helianthus annuus*

Eastern Oregon and as cultivated

Wild sunflower is common in many parts of eastern Oregon and is visited freely by bees. Cultivated sunflower has nectar sugars of 32%.

Sweet Clover, White
Melilotus alba Desr.
LEGUMINOSAE
Annual or biennial
Late June into August
White
n p

Primarily central and eastern Oregon ditches and roadsides

A premier honey plant where present in sufficient abundance.

Fig. 223 *Melilotus alba*

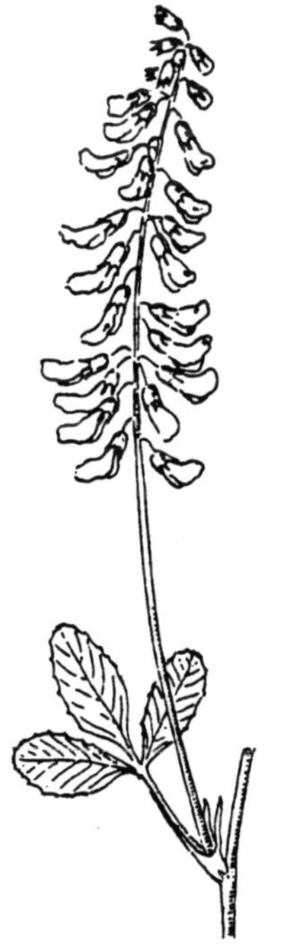

Sweet Clover, Yellow
Melilotus officinalis (L.) Pall.
LEGUMINOSAE
Annual or biennial
June and July
Yellow
n p

Mainly northeastern Oregon

A poorer honey producer and better pollen source than white sweet clover, and blooms about two weeks earlier. Not as widespread as the white species.

Fig. 224 *Melilotus officinalis*

Tamarisk
Tamarix spp.
TAMARICACEAE
Tall spreading bush
Spring
Pink
n p

Fig. 225 *Tamarix parviflora*; a, pistillate flower;
b, staminate flower; c, catkins; d, leaves.

As planted

Very hardy plants which are occasionally planted as ornamental trees. Too few in number to be of much value. Also called salt cedar.

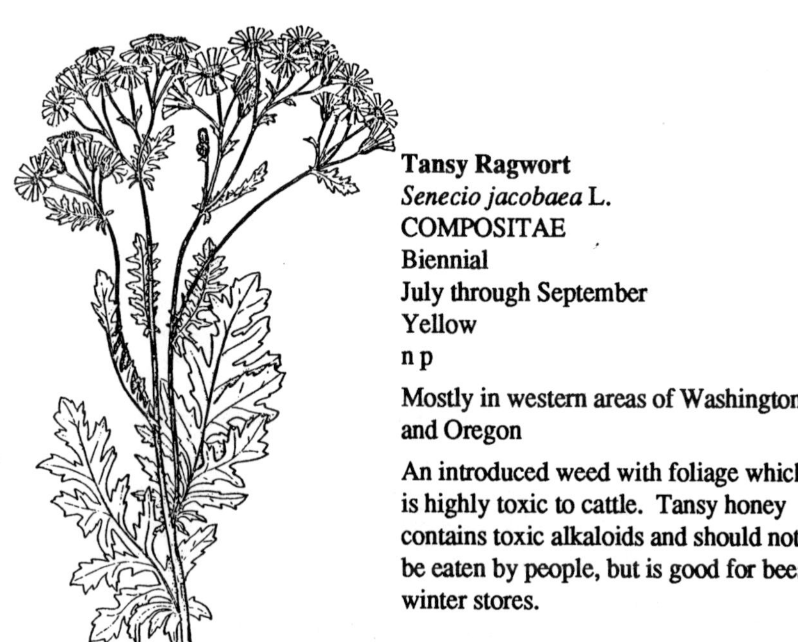

Tansy Ragwort
Senecio jacobaea L.
COMPOSITAE
Biennial
July through September
Yellow
n p

Mostly in western areas of Washington and Oregon

An introduced weed with foliage which is highly toxic to cattle. Tansy honey contains toxic alkaloids and should not be eaten by people, but is good for bees winter stores.

Fig. 226 *Senecio jacobaea*

120

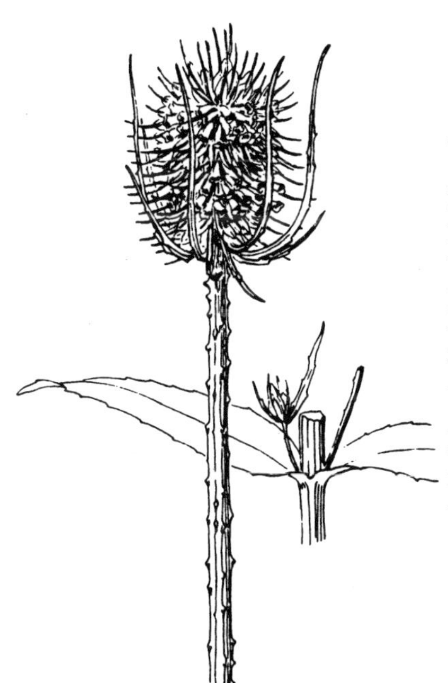

Teasel
Dipsacus sylvestris Huds.
DIPSACACEAE
Large biennial
Summer
Lavender
n p

Widespread weed in moist habitats

Native of Europe. Common along old river beds and railways. Very attractive to bees.

Fig. 227 *Dipsacus sylvestris*

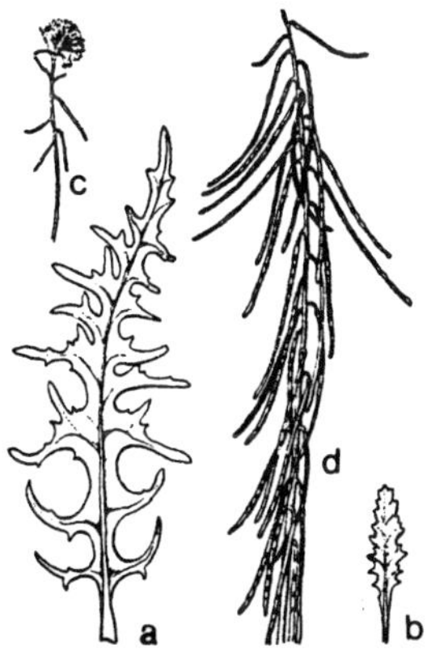

Thelypodium
Thelypodium spp.
CRUCIFERAE
Annual, biennial or perennial
Summer
White, yellow
n ?

Fig. 228 *Thelypodium lasiophyllum*; a, b, types of basal leaves; c, flower; d, fruiting branch.

Eastern Oregon

A species of this genus is reported as a source of nectar in central Oregon.

Thimbleberry
Rubus parviflorus Nutt.
ROSACEAE
Shrub
May and June
White
n p

Fig. 229 *Rubus parviflorus*

General

Not of much importance because of low flower density.

Thistle, Bull
Cirsium vulgare (Savi) Tenore
COMPOSITAE
Biennial
Mid to late summer
Rose
n p

Widespread as an introduced weed

Honey bees work it freely. Nectar sugars at 38%. A native of Eurasia.

Fig. 230 *Cirsium vulgare*; a, flowering
stem, b, leaf.

Fig. 231 *Cirsium arvense*; a, staminate plant;
b, pistillate plant.

Thistle, Canada
Cirsium arvense (L.) Scop.
COMPOSITAE
Perennial
Early summer
Purple
n p

Generally in fields and river bottoms

An introduced Eurasian weed that rapidly became general in distribution. An important source of nectar and pollen where common. Nectar sugars range between 34-47%. Honey mild and white. Male and female flowers on separate plants.

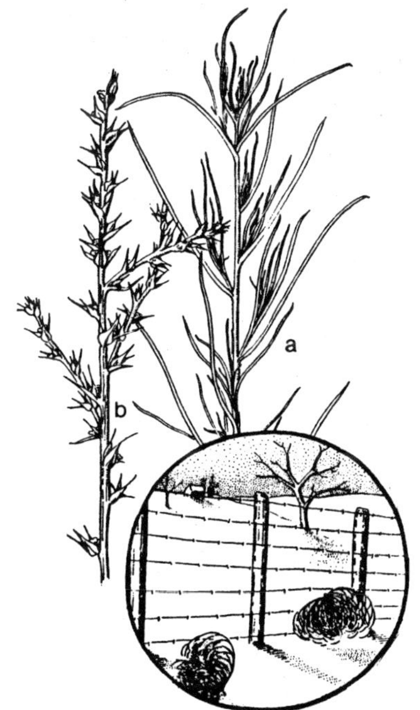

Fig. 232 *Salsola kali*; a, spring, new leaves;
b, summer, leaves and flowers.

Thistle, Russian
Salsola kali L. var. *tenuifolia* Tausch.
CHENOPODIACEAE
Annual
June to August
Green
n p

Chiefly in eastern Oregon

An introduced weed from Europe. Commonly called tumbleweed. Inconspicuous flowers followed by pink ruffled fruits.

Fig. 233 *Sonchus oleraceus*; Annual
sowthistle; a, basal leaf; b, fruit.

Thistle, Sow
Sonchus spp.
COMPOSITAE
Annual and perennial herb
Early summer
Yellow
n p

General

Introduced weed from Europe. Bees
seldom seen on it in Oregon.

Fig. 234 *Hypericum anagalloides*

Tinker's Penny
Hypericum anagalloides C. and
S.
HYPERICACEAE
Annual
June to July
Yellow
n p

General in moist ground

Also called Water St. John's-Wort or Bog St. John's-Wort. Nectar sugar
concentration 48-60%. Pollen is orange-colored.

Toad Flax
Linaria vulgaris Hill
SCROPHULARIACEAE
Perennial
June to September
Yellow
n p

Weedy; general

Also called butter-and-eggs. Honey bees take nectar from the deep blossoms through openings made by bumble bees. The plant is too uncommon to be of much value.

Fig. 235 *Linaria vulgaris*

Tree of Heaven
Ailanthus altissima
(Mill.)Swingle
SIMAROUBACEAE
Tree
Summer
Greenish white
n p

Fig. 236 *Ailanthus altissima*; a, leaves; b, flower; c, winged fruit.

As planted

This Chinese tree is planted as a shade tree and highway tree in Eastern Oregon. Separate male and female trees. Nectar is produced by conspicuous glands on the leaves.

Trefoil
Lotus spp.
LEGUMINOSAE
Annual or perennial
Late spring and summer
Yellow

n p

General

Birdsfoot trefoil, *L. corniculatus* L., may be a secondary source in irrigated pasture in Rogue Valley. *Lotus major* may be locally important on south coast. Reports of bees being very irritable when working this crop.

Fig. 237 *Lotus corniculatus*; a, seed pods.

Fig. 238 *Liriodendron tulipifera*; a, flowering branchlet; b, fruit.

Tulip Tree, Tulip Poplar
Liriodendron tulipifera L.
MAGNOLIACEAE

Tree
Late May and early June
Greenish yellow   n p

As planted in Western Oregon

Occasionally planted for shade in the Northwest, this tree is an important surplus-producing plant of the southeastern states. It is too uncommon to be of much value.

Verbena
Verbena spp.
VERBENACEAE
Perennial
Summer
Blue, white
n -

Open moist ground throughout the state

Some species have been reported as important nectar sources. The most common species in Oregon, *V. hastata* L., occurs too sparsely to be of much value.

Fig. 239 *Verbena hastata*; Tall wild verbena

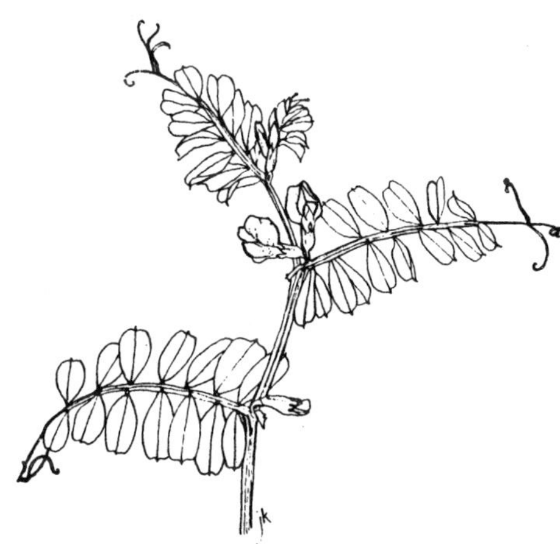

Vetch, Common
Vicia sativa L.
LEGUMINOSAE
Perennial
Summer
Purple
- -

Fig. 240 *Vicia sativa*

General

An extremely variable species, introduced from Europe and now widespread. Flowers produced in leaf axils compared with the racemes of hairy vetch.

Vetch, Hairy
Vicia villosa Roth
LEGUMINOSAE
Annual
Late May through June
Purple
n p

Commercially planted in northern Willamette Valley, and as an escape

A premier honey plant and a major source in the Rogue Valley. Acreage and importance have declined in Willamette Valley. Source of light, mild honey. Nectar sugar concentration about 45%.

Fig. 241 *Vicia villosa*

Virginia Creeper
Parthenocissus quinquefolia
(L.) Planch.
VITACEAE
Climbing vine
July and early August
Cream
n p

Fig. 242 *Parthenocissus quinquefolia*

As planted

Very attractive to bees for both pollen and nectar, but not common enough to be of much value.

Fig. 243 *Vitex negundo*

Vitex
Vitex negundo L.
VERBENACEAE
Shrub or small tree
Summer
Blue, white
n -

As planted

Specimens were planted by beekeepers in Oregon in the 1930's but it was found that Vitex often winter-killed here. Eastern US beekeepers report it as a valuable nectar source.

Fig. 244 *Juglans regia*

Walnut
Juglans spp.
JUGLANDACEAE
Tree
Spring
Yellow catkins
- p

As planted, mostly in western Oregon

Considerable pollen is produced by both the English and Black walnut. "Oak-dew" may also be produced.

Fig. 245 *Citrullus lanatus*; a, male flower; b, female flower.

Watermelon   Annual vine
Citrullus lanatus (Thunb.) M.& N.   Summer
CUCURBITACEAE   Yellow   n p

Mainly Umatilla County

Commercially grown in northeastern Oregon along the Columbia River. Honey bee pollination greatly benefits fruit set, size and shape.

Whitlow Grass
Draba verna L.
CRUCIFERAE
Annual
February to May
White
n p

General

Reported as yielding nectar and pollen in the Hermiston area.

Fig. 246 *Draba verna*

130

Fig. 247 a, *Salix scouleri*; b, *S. sitchensis*; c, pistillate flower; d, staminate flower.

Willow   Shrub and tree
Salix spp.   February and March
SALICACEAE   Greenish yellow catkins   n p

General in moist soil

The many species of willow are often the earliest flowers of major importance for spring buildup. Greenish yellow catkins furnish both pollen and nectar. Sugars in nectar may be in excess of 50%. Willows are dioecious plants.

Wisteria
Wisteria spp.
LEGUMINOSAE
Perennial vine
April and May
White, blues, pink
n -

Fig. 248 *Wisteria sinensis*

As planted

Very attractive to bees, but too uncommon to be very significant.

Fig. 249 *Wyethia angustifolia*

Wyethia
Wyethia spp.
COMPOSITAE
Low perennial herb
Spring
Yellow
n p

Throughout the state

Also called wild or prairie sunflower or Mule's-ears. It has yellow flowers and large leaves. Some nectar and pollen produced by *W. angustifolia* (DC.) Nutt. in the west and *W. amplexicaulis* Nutt. in the east.

Fig. 250 *Eriodictyon californicum*;
a, flowering branchlet; b, leaf.

Yerba Santa
Eriodictyon californicum (H.& A.)Torr.
HYDROPHYLLACEAE
Shrub
Early summer
White, pale blue
n p

Southern Oregon

In hills of California this plant is valuable to bees and produces a spicy flavored honey with reputed cathartic properties. Its value to bees in Oregon has not been determined.

FLOWER COLORS

White, cream
Abelia
Apple
Apricot
Aster
Bachelor's Button
Basswood
Bean
Blackberry
Buckeye
Buckwheat
Burnet
Cactus
Carrot
Catalpa
Catnip
Ceanothus
Chaparral Broom
Cherry
Chestnut, Horse
Chickweed
Chinquapin
Clematis
Clover, Arrowleaf
Clover, White
Corn
Cotoneaster
Crabapple
Crocus
Daisy, Michaelmas
Death Camas
Dodder
Dogbane
Dogwood
Elderberry
Eriogonum
Everlasting, Pearly

False Hellebore
Fawn Lily
Foxglove
Hawthorn
Heath and Heather
Holly
Hollyhock
Horehound
Hyssop
Indian Peach
Labrador Tea
Laurel, English
Laurestinus
Lilac, Wild
Loco Weed
Locust, Black
Loganberries
Lupine
Madrone
Mallow
Manzanita
Marigold
Meadowfoam
Miner's lettuce
Mock Orange
Morning Glory
Mustard, Jim Hill
Nightshade
Ninebark
Ocean Spray
Old-man-in-the-ground
Onion
Parsnip, Cow
Pear
Penstemon

Phacelia
Photinia
Plagiobothrys
Plantain, English
Plums and Prunes
Poison Ivy
Polygonum
Potentilla
Privet
Psoralea
Pyracantha
Quince
Raspberry
Rhododendron
Sage
Salal
Service Berry
Sidalcea
Silene
Snowberry
Soap Root
Sorrel, Wood
Spearmint
Spiraea
Starwort
Strawberry
Sweet Clover
Thelypodium
Thimbleberry
Verbena
Virginia Creeper
Vitex
Whitlow Grass
Wisteria
Yerba Santa

Yellow, orange

Antelope Brush
Asparagus
Balsam Root
Barberry
Beggar-ticks
Buttercup
Cabbage
Cactus
Chaparral Broom
Cleome
Clover, Bur
Clover, Small Hop
Crocus
Cucumber
Dandelion
Dandelion, False
Evening Primrose
Fawn Lily
Fiddleneck
Goldenrod
Goldfield
Greasewood
Grindelia
Hollyhock
Honeysuckle
Knapweed
Lettuce
Lupine
Marigold
Milkweed
Mullein
Muskmelon
Mustard
Oregon Grape
Parsnip
Pea
Phacelia
Poppy, California
Potentilla
Pumpkin, Squash, Gourd
Rabbit Brush
Radish
Rape
Rhododendron
Russian Olive
Sagebrush
Scots Broom
Skunk Bush
Skunk Cabbage
Sneezeweed
Sorrel, Sheep
Sorrel, Wood
Spikeweed
Spring Gold
St. John's-Wort
Star Thistle
Sunflower
Sweet Clover
Tansy Ragwort
Thelypodium
Thistle, Sow
Tinker's Penny
Toad Flax
Trefoil
Watermelon
Wyethia

Pink, red, rose

Abelia
Apple
Aster
Bachelor's Button
Bean
Bindweed
Blackberry
Ceanothus
Chestnut, Horse
Clover, Arrowleaf
Clover, Crimson
Clover, Red
Cotoneaster
Currant
Daisy, Michaelmas
Dogbane
Fawn Lily
Filaree
Fireweed
Hawthorn
Heath and Heather
Hedge-nettle
Hollyhock
Honeysuckle
Huckleberry
Laurel, Pale
Locust, Black
Mallow
Manzanita
Maple, Vine
Milkweed
Miner's Lettuce
Onion
Pea
Pennyroyal
Peach
Polygonum
Pussy Paws
Quince
Red Maids
Rhododendron
Rose
Sainfoin
Salal
Salmonberry
Sidalcea
Silene
Snowberry
Sorrel, Sheep
Spiraea
Tamarisk
Thistle, Bull
Wisteria

Blue, purple, lavender

Alfalfa
Aster
Bachelor's Button
Bluebell
Blue Curls
Borage
Burdock
Burnet
Catnip
Chicory
Clover, Strawberry
Clover, Three Toothed
Cone Flower
Crocus
Daisy, Michaelmas
Dead-nettle, Red
Downingia
Farewell-to-Spring
Figwort
Filaree
Fireweed
Flax
Foxglove
Heal-all
Heath and Heather
Hedge-nettle
Hound's Tongue
Hyssop
Knapweed
Laurel, Pale
Lettuce
Lilac, Wild
Loco Weed
Lupine
Milkweed
Morning Glory
Nightshade
Pea
Pennyroyal
Penstemon
Peppermint
Phacelia
Potentilla
Psoralea
Rhododendron
Rocky Mountain Bee
Plant
Sage
Sidalcea
Sorrel, Wood
Spearmint
Teasel
Thistle, Canada
Verbena
Vetch
Vitex
Wisteria
Yerba Santa

Green, greenish

Ash
Box Elder
Burnet
Cascara
Celery
Cocklebur
Coffee Berry
Cottonwood
Elm
Grape
Grass
Huckleberry
Ivy, Boston
Ivy, English
Laurel, California
Locust, Honey
Mahogany, Mountain
Maple, Big-leaf
Maple, Norway
Mistletoe
Mulberry
Mullein, Turkey
Mustard, Tansy
Parsley
Poison Oak
Polygonum
Sumac
Thistle, Russian
Tree of Heaven
Tulip Tree

Catkins or cones

Alder
Aspen
Birch
Cedar
Cottonwood
Filbert
Fir
Hazelnut
Oak
Pine
Spruce
Walnut
Willow

BIBLIOGRAPHY

Abrams, LeRoy and Roxana Stinchfield Ferris. (1960) Illustrated Flora of the Pacific States, 4 Volumes. Stanford University Press, Stanford, California.

Bailey, L. H. (1947) The Standard Cyclopedia of Horticulture. Macmillan Company, New York, New York.

Crane, E., Walker, P. and R. Day. (1984) Directory of Important World Honey Sources. International Bee Research Association, Cardiff, Wales, UK.

Dadant and Sons, ed. (1975) The Hive and the Honey Bee. Dadant and Sons, Hamilton, Illinois.

Dennis, La Rea J. (1980) Gilkey's Weeds of the Pacific Northwest. Oregon State University Bookstores, Inc., Corvallis, Oregon.

Franklin, Jerry F. and C.T. Dyrness. (1973) Natural Vegetation of Oregon and Washington. USDA Forest Service General Technical Report PNW-8

Gilkey, Helen M. and La Rea J. Dennis. (1980) Handbook of Northwestern Plants. Oregon State University Bookstores, Inc., Corvallis, Oregon.

Grant, John A. and Carol L. Grant. (1943) Trees and Shrubs for Pacific Northwest Gardens. Dogwood Press, Seattle, Washington.

Hayes, Doris W. and George A. Garrison. (1960) Key to Important Woody Plants of Eastern Oregon and Washington. United States Department of Agriculture, Forest Service; Agriculture Handbook No. 148.

Hitchcock, C. Leo and Arthur Cronquist. (1973) Flora of the Pacific Northwest. University of Washington Press, Seattle, Washington.

Hitchcock, C. Leo and others. (1955-1969) Vascular Plants of the Pacific Northwest, 5 parts. University of Washington Press, Seattle, Washington.

Hooper, Ted and Mike Taylor. (1988) The Beekeeper's Garden. Alphabooks Ltd, A&C Black, London, UK.

Howes, F.N. (1979) Plants and Beekeeping. Faber and Faber Ltd, London and Boston.

Jepson, Willis Linn. (1925) A Manual of Flowering Plants of California. University of California Press, Berkeley, California.

Kruckeberg, Arthur. R. (1982) Gardening with Native Plants of the Pacific Northwest. University of Washington Press, Seattle, Washington.

Krussmann, Gerd. (1976) Manual of Cultivated Broad-leaved Trees and Shrubs. Timber Press, Inc., Portland, Oregon.

Labadie, Emile L. (1980) Ornamental Shrubs for use in the Western Landscape. Sierra City Press, Sierra City, California.

Liberty Hyde Bailey Hortorium. (1976) Hortus Third, Concise Dictionary of Plants Cultivated in the United States and Canada. Cornell University, Ithaca, New York.

Lovell, Harvey. (1966) Honey Plants Manual. A.I. Root Co., Medina, Ohio.

Lovell, John H. (1926) Honey Plants of North America. A.I. Root, Co., Medina, Ohio.

McGregor, S.E. (1976) Insect Pollination of Cultivated Crop Plants. Agricultural Handbook No. 496, United States Department of Agriculture, Washington, DC.

Martel, D.J. and G.N. Fredeen. (1981) Plant Materials for Landscaping; A List of Plants for the Pacific Northwest. Pacific Northwest Cooperative Extension Publication PNW 185.

Nye, W.P. (1971) Nectar and Pollen Plants of Utah. Monograph Series XVIII (3). Utah State University Press, Logan, Utah.

Pellett, Frank. C. (1977) American Honey Plants. Dadant and Sons, Hamilton, Illinois.

Ramsay, Jane. (1987) Plants for Beekeeping in Canada and the Northern USA. International Bee Research Association, Cardiff, Wales, UK.

Rickett, Harold William. (1971) Wild Flowers of the United States: Vol. 5, The Northwestern States. The New York Botanical Garden, McGraw-Hill Book Company, New York, New York.

Root, A.I. and others. (1983) ABC and XYZ of Bee Culture. A.I. Root, Co., Medina, Ohio.

Sargent, Charles S. (1949) Manual of Trees of North America. Dover Books, New York, New York.

Scullen, H.A. and G.A.Vansell. (1942) Nectar and Pollen Plants of Oregon. Oregon Experiment Station Bulletin 412, Oregon State College, Corvallis, Oregon.

Sudworth, George B. (1908) Forest Trees of the Pacific Slope. United States Department of Agriculture, Forest Service.

Vansell, G.A. (1941) Nectar and Pollen Plants of California. Bulletin 517, University of California, Berkeley, California.

ABOUT THE AUTHORS

Michael Burgett is an associate professor in the Department of Entomology at Oregon State University. He is the director of the Honey Bee Laboratory at OSU and is responsible for teaching, research and extension efforts in apiculture at OSU.

La Rea Johnston is the Assistant Curator of the Oregon State University Herbarium. She has taught plant taxonomy classes for 30 years and was recently awarded the Burlington Northern Foundation Faculty Achievement Award for excellence in teaching. Mrs. Johnston identifies plant specimens for the Agricultural Extension Service and anyone else in the state.

Beryl Stringer is a beekeeper and Chairperson of the Nectar and Pollen Plants Committee of the Oregon State Beekeepers Association. She has published articles about plants and bees in *National Gardening Magazine, Gleanings in Bee Culture, American Bee Journal* and the OSBA newsletter, *The Bee Line.*

Should any reader have further observations concerning nectar and pollen plants in the Pacific Northwest, we would appreciate such comments being sent to the authors at Honeystone Press, PO Box 511, Blodgett, Oregon 97326.